ADVANCE PRAISE FOR

Pathology & Technology

"How did 'sick' become the new normal? How do the stories we tell about disordered, degenerate, and abnormal media users reveal reflections of ourselves? From anthrax to viruses, from neurasthenia to schizophrenia, technopathologies, in D. Travers Scott's able hands, reveal more than technical conditions: they reveal our many humanities. Drawing together diverse cultural, sociological, political, and critical theoretical sources, this lively cultural historical account enriches scholarly understanding of how modern communication, and its makers, police and pathologize the technical limits of human behavior. Confident in the face of critical difference and evenhanded in its analysis, Scott has opened a new chapter on the medical-moral panics, the scapegoating thrills, and unease we rediscover in every new disease. A vital read."

—Ben Peters, Associate Professor of Media Studies,
Henry Kendall College of Arts & Sciences, University of Tulsa

"*Pathology & Technology* is a fresh and original book—a deeply researched study of how (some) tech users are demonized as diseased. D. Travers Scott traces the media and popular discourses that label some technologies—or really their users—as 'sick'. Mixing history, focus group interviews, and discourse analysis, the book is a rich investigation of how 'technopathologies' emerge and circulate. *Pathology & Technology* is ultimately a book about invisible politics—about how medicalized tech talk renders and then contains 'bad users.'"

—Jeff Pooley, Muhlenberg College; Author of *James W. Carey and Communication Research: Reputation at the University's Margins*

"Eloquent and incisive, D. Travers Scott's *Pathology & Technology: Killer Apps & Sick Users* examines the complex history of pathologizing discourses surrounding new technologies. His historically grounded, theoretically nimble study suggests that our current obsession with technology-generated sicknesses may reveal more about our cultural anxieties surrounding gender, sexuality, and power than technology or illness. Provocative and ground-breaking, this project reframes questions of technology, illness, and agency in a productive and compelling fashion."

—Jennifer Natalya Fink, Associate Professor,
Department of English, Georgetown University

"The idea that communication technology can be bad for us is a well-worn groove in Western culture, one whose invocation can be so expected that we fail to note when it happens. With a historian's flair for the telling detail, D. Travers Scott's *Pathology & Technology* expertly sidesteps the traps awaiting anyone traversing the history of communication technology and its invitation to all order of determinisms and faulty assumptions. Scott relies on historical evidence and direct engagement with persons to present a narrative not about technologies per se, but about disease discourses as they have been applied to technology. This requires some tight methodological and theoretical maneuvering, and Scott is up to the task. The result is a book that accomplishes something remarkable: *Pathology & Technology* is a definitive, user-centered history of how pathologization comes to order our understanding of communication technology."

—David W. Park, Professor of Communication,
Lake Forest College

Pathology & Technology

This book is part of the Peter Lang Media and Communication list.
Every volume is peer reviewed and meets
the highest quality standards for content and production.

PETER LANG
New York • Bern • Berlin
Brussels • Vienna • Oxford • Warsaw

D. Travers Scott

Pathology & Technology

Killer Apps & Sick Users

PETER LANG
New York • Bern • Berlin
Brussels • Vienna • Oxford • Warsaw

Library of Congress Cataloging-in-Publication Data

Names: Scott, D. Travers, author.
Title: Pathology and technology: killer apps and sick users / D. Travers Scott.
Description: New York: Peter Lang, 2018.
Includes bibliographical references and index.
Identifiers: LCCN 2017038860 | ISBN 978-1-4331-4845-3 (pbk.: alk. paper)
ISBN 978-1-4331-4846-0 (hardback: alk. paper) | ISBN 978-1-4331-4847-7 (ebook pdf)
ISBN 978-1-4331-4848-4 (epub) | ISBN 978-1-4331-4849-1 (mobi)
Subjects: LCSH: Environmental health.
Technology—Risk assessment.
Health risk assessment.
Technology—Health aspects.
Classification: LCC RA565 .S37 2018 | DDC 616.9/8—dc23
LC record available at https://lccn.loc.gov/2017038860
DOI 10.3726/b11843

Bibliographic information published by **Die Deutsche Nationalbibliothek**.
Die Deutsche Nationalbibliothek lists this publication in the "Deutsche Nationalbibliografie"; detailed bibliographic data are available on the Internet at http://dnb.d-nb.de/.

29 Broadway, 18th floor, New York, NY 10006
www.peterlang.com

For Sarah Banet-Weiser

TABLE OF CONTENTS

ILLUSTRATIONS

ACKNOWLEDGMENTS

I am deeply grateful to the many individuals who contributed to and supported this project. Apologies to anyone unnamed in this surely incomplete list: Anne Balsamo, Sarah Banet-Weiser, my advisor and committee chair; Manuel Castells, Sasha Costanza-Chock, Madeline Crowley, Joanna Demers, Nina Eliasoph, Rosa Linda Fregoso, Jennifer Natalya Fink, Alice Gambrell, G. Thomas Goodnight, Herman Gray, Larry Gross, Raquel Gutierrez, Velina Hasu Houston, Aniko Imre, Ralina Joseph, Josh Kun, Erin Maher, Mike McGirr, Tara McPherson, Roopali Muhkerjee, Sheila Murphy, Kent Ono, Lisa Parks, Steven Rafferty, Michael Schudson, Vanessa Schwartz, David Silver, Matthew Swank; Doug Thomas, Crispin Thurlow, Allison Trope, Karina Wilson, Genevieve Yue, anonymous reviewers, my cohort and colleagues at the University for Southern California Annenberg School for Communication and Journalism and the Clemson University Department of Communication, my focus group participants and interview subjects, the staff at House of Marketing Research, Pasadena, CA; the staff and librarians of the Huntington Library and Archives, the Margaret Herrick Library of the Academy of Motion Picture Arts and Sciences, the ONE Institute and Archives, the UCLA Film and Television Archive, the U.S. Library of Congress, and the USC Libraries and Special Collections; and the reviewers, panelists, respondents, and audience members of various conference presentations. Research for this project was partially funded by a 2008–09 New Directions in Feminist Research Fellowship from the USC Center for Feminist Research, a 2009–10 Dibner History of Science Fellowship from the Huntington Library and Archives, and 2007 and 2009 Stark Family Fellowships for Summer Research from USC Annenberg School for Communication and Journalism. Portions of this manuscript appeared in different forms in *American Quarterly*, *Angelaki*, *Feminist Media Studies*, and *Rocky Mountain Communication Review*.

INTRODUCTION

Pathological Technoculture: Sick Users and Reinforced Stereotypes

The Pain of Staying Connected: Experts Are Seeing an Increase in Stress Injuries from Overuse of Hand-Held Gadgets

—*ABC News*, 2007

Cancer Risks Seen in Use of Cellphone

—Charleston, SC, *Post & Courier*, 2011

The E-Ego: The Web Brings Out the Worst in Us

—*Psychology Today*, 2014

Instagram Worst Social Media App for Young People's Mental Health

—*CNN*, 2017

Such headlines are common. In response, sociologists tend to dismiss them as irrational fears of technological progress, conspiracy paranoia, or victim mentalities.[1] Others roll eyes at media sensationalism. Consumers debate and scientists research if they are "real" diseases.

This book, the result of a 10-year research project, will show that what is not in the least ambiguous is the reality of such discourses. Throughout the Western history of electric communication technologies, such devices have been associated with causing or worsening various forms of mental and physical distress. I call these cases *technopathologies*. In addition to documenting and making legible these phenomena and the broader *pathological technoculture* they represent, the animating question that drives this book is: What can we learn by listening to these discourses? Instead of interpreting what they mean or judging whether they are "real," this project aims to listen to what they have to say. Technopathological discourses contribute to ideals and norms of technology users as a subject position, I will argue, one that links them to existing social hierarchies and also attempts to preemptively resist certain challenges to those hierarchies.

Health is not external to social reality. As scholars such as Deborah Lupton have shown, medical science can be approached as an element of culture. By taking an approach of cultural analysis, I do not mean approaching disease discourses as primarily symbolic—as displaced anxieties about cultural change, newness, or evolving social identities. Nor do I reject their metaphoric meanings, as Susan Sontag famously called for in *Illness as Metaphor*. For, as Michael Taussig has described, ignoring the signifying dimensions of diseases mutes the contradictions they express and supports the ideological status quo.[2] Although influenced by anthropologist Mary Douglas's famous work on pollution taboos, I do not solely analyze diseased bodies as symbols of diseased social bodies or metaphors of transgressed modernist boundaries. My aim instead is to take seriously the explicit meanings of a sick user as a sick user, to identify technopathological discourses and examine their entanglements of matter and meanings. I examine disease discourses as activities in which the symbolic and the "real" intertwine and co-create. To use the language of feminist science and technology scholar Karen Barad, they are material-discursive practices, phenomena entangled in their iterations of (re)becoming. Several scholars have identified similar processes in cultural histories of disease.[3] Indeed, this process of pathologizing technology users can be seen as part of the larger medicalization of technology: the territorialization of a social realm into that of medical knowledge and discourses.[4] Medicalization has been explored, for example, in topics such as patriarchal medical power overtaking the previously domestic realms of midwifery and home childbirth,[5] deviance,[6] and rescue missions.[7] Michel Foucault's work on the history of madness and psychiatry is a paradigmatic and formative example of medicalization. As he wrote, "Nineteenth-century psychiatry was a medical science as much for the social body as the individual soul."[8] Medicalization is perhaps inevitable. Our physical and mental vulnerability, as presages of our ultimate mortality, are central to human experience.

Yet, it is nevertheless an informative analytic. By examining how technology use has been discussed pathologically, what insights might we gain into the evolving ideals of humans and technologies, in our roles as users? How have these discourses changed over time? If we examine them across various media, institutions, and sites, what patterns and resonances might emerge?

While extending and applying scholarship on technology and the body in general,[9] I focus specifically on bodily insights into technological practices. Carolyn Thomas de la Peña argues that Americans' early-twentieth-century enthusiasm for electricity, radium, and energy shows how new technologies were understood through bodily contexts.[10] Carolyn Marvin likewise argues that initial understandings of electrification were based on bodily references and applications. However, I wish to move beyond any direction of a causal arrow between technologies and bodies. Instead, I will examine cases in which the two are combined, in which bodies and technological practice intersect: cases of users.

My emphasis on the bodies and minds of users is not a focus on bodies affected by technology. Bodies are not combined, fused, prostheticized, conglomorated, transcended, duplicated, extended, informatized, uploaded, or transformed. They are simply bodies. This is not to dismiss technology; it is to offer a perspective that rejects the ontological separation of technologies and bodies as disparate things. They have been endlessly intertwined. Moreover, as this book demonstrates, technopathologies are not new, nor are they primarily about new media or innovations. Part of the perspective with which I argue this is one that does not see technologized bodies as radically different. Instead, I will argue, we have always been users. Furthermore, as I will show, technopathologies have a deep-rooted logic. Associations of biological and technological phenomena are neither fantastical nor irrational, but based in a long scientific and philosophical history, which I will examine later in a discussion of electricity.

This project intervenes in the presumed linkage of technological "newness" and utopian and dystopian cultural responses. This relates to what I refer to as the *newness hypothesis*: assertions that technopathologies are merely technophobic reactions to new media by irrational neo-Luddites, or, at best, unsophisticated naïfs. The newness hypothesis suggests that people fear social change, and they project these fears onto new media as symbols of such change. While not entirely untrue, this is far from the whole story. I will show that technopathologies have also often occurred with well-domesticated, dominant media, and even technologies that are safely residual or "old" media. The historical persistence of technopathologies pushes us to hear the ongoing processes and entanglements involved in usership. This is in contrast to studying technology using linear relations of cause and effect, which reinforce technological determinism and separate the realms of technology and culture. Furthermore, new technologies do not always provoke technophobic reactions of the sort with which technopathologies are often grouped. For example, in contrast to existing histories of awe, fear, and excitement around the onset of electrification, I will show how, in the case of Anderson, South Carolina—"The Electric City"—newness could also be experienced as a very level-headed "electrical banal." Technopathological discourses are about users, not technologies. Pathologization focuses on individual, rather than systemic or social concerns. In these discourses, users are aligned in many ways with existing social hierarchies, such as those of age, gender, race, sexuality, and mental health. This expands sick users to a broader conception of bad users. Technopathologies are social diseases, and it is the user's responsibility to manage them. When users get sick, they begin to pose a threat to other users. As with citizens who are healthy and sick, or good and bad, users have provoked a variety of cultural and legislative responses, and ultimately, sickness becomes a regular part of technology use. In this *normalization of the abnormal*, disease becomes an accepted and familiar part of usership.

What Are Technopathologies?

The cacophony of stimuli short-circuits both conscious and unconscious thought, preventing our minds from thinking either deeply or creatively. Our brains turn into simple signal-processing units, shepherding information into consciousness and then back out again.

> —"How the Internet is Making Us Stupid," Nicholas Carr's *Telegraph* article on the research behind his book, *The Shallows: How the Internet Is Changing the Way We Think, Read and Remember*, a finalist for the 2011 Pulitzer Prize in General Nonfiction.

Such perspectives of technologies inducing deleterious effects on individual minds and bodies are common in news, entertainment, and the classroom. Friends, family, and colleagues have confidently paraphrased them to me across dinner tables and in conference rooms. Indeed, one can readily assembly a medical encyclopedia of technological ailments from recent news articles:

- "Google makes us all dumber"[11]
- "iPhone Separation Anxiety Makes You Dumber"[12]

- "Depression, loneliness linked to binge-watching TV"[13]
- "E-books 'damage sleep and health'"[14]
- "Man Tears Tendon After Playing Candy Crush for Weeks"[15]
- "Casual Sex Apps Blamed for Skyrocketing STD Rates"[16]
- "Touchscreen use linked to sleep problems in infants"[17]
- "Temporary blindness tied to smartphone use in dark"[18]
- "Too much Facebook can kill you"—when you ignore your girlfriend and drive her to murder you[19]
- "Mobile phones are cooking men's sperm"[20]
- "Phubbing"—using a cellphone during dinner—can ruin interpersonal relationships[21]
- "Selfies lead to 'social media lice'"[22]
- "Brain scramble," from multitasking on computers and mobile devices, disrupts sleep, memory, intelligence, concentration, circadian rhythms, physical and mental health, and parenting[23]
- "Computer Face," a permanent expression of frowning, wrinkles, and double necks[24]
- "Computer Vision Syndrome," a state of pain, discomfort, blurred vision and headaches[25]
- Virtual reality has a "vomit problem"[26]
- Anticipation of replacing devices results in an "Upgrade Effect," the irrational abuse of current devices[27]
- "Screen Addiction" occurs among children[28] and in the "secret lives" of teens[29]
- Smartphones cause headaches and "occipital neuralgia," a form of nerve damage[30]
- Video games distort our sense of time[31]
- The iPhone 6 can rip hair off your face[32]
- Immersive and interactive games like Pokémon Go blur reality in young users' minds[33]
- Mobile phones and apps have caused increased nearsightedness, changes in brain functions, dopamine-fueled "compulsion loops," nomophobia—the fear of being without your phone, and disrupted sleep patterns (leading to diabetes, cancer, and obesity)[34]
- Selfies kill more people annually than Mt Everest,[35] shark attacks,[36] or American football, and are most prevalent in India, to such a degree that "no-selfie zones" have been established[37]
- Portable music players cause hearing loss[38]

This book explores such intersections between cultural fields of medicine and technology under the rubric of *pathological technoculture*. Technoculture is a term from critical technology studies that conceives of technologies as wholly embedded within the social system of the cultures that produce, use, and shape them, rather than as discrete causal agents impacting culture as if from outside. Pathological literally means related to disease. I employ it here to include also the social process of pathologization. If medicalization is the process by which various non-medical elements of social reality become territorialized by or subsumed under medical authority, pathologization is a subset of this phenomenon specifically focused on disease. Pathologization is the process by which various social phenomena are not only medicalized, but also explicitly associated with or as disease. For example, an impatient, fidgety disposition was seen as an inherent flaw of character during my childhood, but psychiatry medicalized this as anxiety, possibly caused by trauma, and thereby not a fault of the subject. Neuropsychology pathologized this as a neuro-chemical imbalance, the disease of Attention Deficit Hyperactivity Disorder.

Technopathologies represent the pathologization of technoculture. From news, scholarship, entertainment, advertising, and medical journals, I have gathered hundreds of examples of ill health attributed to electric communication technologies. Although pathologization offers a broad conceptual umbrella, it is difficult to use a more specific term with precision. I do not limit my investigation to official diagnoses in authoritative sources, such as the American Psychiatric Association's *Diagnostic and Statistical Manual of Psychiatric Disorders*, the U.S. National Institutes of Health's *MedlinePlus* website, or *Physicians Desk Reference* of prescription guidelines. For one, this would introduce a Western cultural bias. For example, neurasthenia has been discredited in the West since World War I, but was a valid diagnosis in some parts of Eastern Europe into the 1990s.[39] Second, it would reinforce medical science as the arbiter of truth, and I am following technopathological discourse even as it spreads to fictional and liminal sources. For example, "text neck" is a term coined to describe repetitive stress injuries caused by looking down at a mobile device.[40] You can learn about this "global epidemic" on the website for The Text Neck Institute (http://text-neck.com), in an article in the journal *Surgical Technology International*, download a mobile app to prompt you to keep your head elevated, or watch a segment about it on the daytime TV show, *The Doctors*, but it is not an officially recognized diagnosis or "real disease" in any medical sources.

Ultimately, the reality of diseases is not my point. This project explores the reality of disease discourses. Throughout the examples in this book you will see colloquial and interchangeable use of "disease," "illness," "sickness," "syndrome," and other terms. Generally I will use the same term as my source to attempt to stay true to the discursive artifact.

Historically, there have been (and continue to be) many meanings to the word "disease." Medical philosophy and sociology have argued for differentiation between *disease* as biological aberration, *illness* as the phenomenological experience of being unhealthy, and *sickness* as public performance of the social role of an unhealthy person.[41] However, marking distinctions between words that refer to varying forms of unwellness and typical functioning of mind and body would assert a presentism I wish to avoid. Therefore, I also engage archaic and discredited historic diagnoses. To exclude historic disease patterns would privilege the present, suggesting that our contemporary diagnoses and disciplinary divisions of knowledge are, in contrast, entirely accurate and "real." For example, research topics that today might be considered delusions to be addressed by psychiatry—studies of demon possession, witchcraft, neurasthenia, and mental disease—were, as recently as the late nineteenth and early twentieth century, part of commingled spiritualism and philosophy.[42] Changes in disciplinary divisions and medical discourses about diseases and their etiologies do not necessarily or entirely dis-entangle longstanding knowledges and relations. Although officially debunked, historic disease etiologies, such as demon possession, can still operate culturally as pathological frameworks, marginal forms, and traditional behaviors. Indeed, exorcism is still widely practiced, in cases ranging from the granddaughter of actor Morgan Freeman[43] to a college peer of former Louisiana Governor and Vice-Presidential candidate Bobby Jihndal.[44]

Associations between electric communication technologies and disease are clearest when employing a broad, historically informed definition of "disease," one that takes into consideration any form of mental or physical pain, suffering, or malfunction that degrades health and quality of life. Literally, dis-ease. I use "disease" in this sense, as an umbrella for mental or physical trouble or discomfort.[45] This is an epistemological and methodological strategy in order to perceive pathological discourse across a variety of terrains not officially sanctioned as "real" disease. As Joseph Dumit has shown, there is no tidy demarcation between real and unreal diseases, but, rather, a rich a contested zone of what he refers to as "liminal illnesses." In order to not only examine both realms of real and unreal, but also the zone between, I employ this broad, vernacular understanding of "disease."

Therefore, I include cases that, to contemporary perspectives, might not immediately seem like a disease. One might think of disease strictly along the lines of a microorganism that invades one's body and causes it to function improperly, or a traumatic event or chemical imbalance that causes mental illness. However, consider how the case of distracted, texting drivers resonates with a common disease: Alzheimer's. Although the causal mechanism of Alzheimer's is not fully known, the disease is recognized as abnormal protein clumps and brain cell death causing degraded memory and mental acuity, which can result in accidents, injuries, and death. Similarly, the use of mobile communication devices while driving causes the brain of a user to exhibit degraded mental acuity in the form of competing inputs of sensory information, improperly prioritized, with weakened perceptual capacities and reduced reaction times, which can, in turn, result in accidents, injuries, and death. Although not an infectious disease, many have associated texting while driving with threats to health and well-being. "YOU DRIVE YOU TEXT YOU DIE," reads a bumper magnet I acquired at my AAA auto service center. As I will discuss later, this has been referred to as situational unawareness, a technopathology I will describe in the case of the Metrolink train crash outside of Los Angeles.

Finally, my use of "disease" also reflects a political commitment to queer politics and methodologies by challenging the presumed stability of categories, definitions, and subjects. As Ann Cvetkovich has described in her work on everyday trauma, bestowing definitions in categories such as illness incites a drive for cures and solutions that can occlude more thoughtful and complex analyses. She prefers to call depression, for example, "feeling bad."[46]

Disease is not alone. It is a state brought into being through relationships between entities. Contagion, the spread of disease, has a long and varied etymology, but has consistently referred to notions of connection: between living beings and also various phenomena in the world.[47] Contagion is a form of what Karen Barad calls "entanglement," ways in which phenomena are mutually involved in their ongoing processes of (re)becoming. In technopathologies, we see three types of relational disease states: cause, connection, and association. One, devices can directly cause you to become sick or sicker. Two, devices can provide a necessary connection to other things that make you sick, just as intercourse can connect you to HIV or a piece of raw tuna can connect you to salmonella bacteria. Note how, in both of those cases, we colloquially think of those actors as bad, immoral people and "bad sushi." This suggests the third relationship, that of association, in which communication devices mark or identify sick and sickening persons. A subsequent chapter will explore this further in terms of the historic medical condition of degeneracy.

People experiencing technopathologies are referred to using diverse terminology, including sufferers, victims, and patients. Again, I am not insisting on applying unified distinctions across the diversity of the discursive artifacts. When using such terms myself, I am simply referring to persons experiencing a deleterious effects associated with an electric communication technology, by which I refer primarily to devices powered or assisted by electricity to facilitate the exchange of information or cultural expression between persons. Although one can identify pathologies associated with automotive, engineering, or other technologies, my focus is on electric communication technologies as a fertile nexus for understanding usership. Although I focus on obvious cases, such as electric lights, telegraph, wired and wireless telephones, motion pictures, radio, television, video and audio recording systems, personal computers and online technologies, there will also be cases of technologies that have been less covered by scholarly research, such as laser printers. In short, I am employing the very contemporary and colloquial usage of "technology," as reflected in my artifacts, although I will address at later points deviations from and exceptions to this. I am not, however, making an argument about any kind of philosophical notion of an essential, generalized "technology" that shapes humanity, society, or consciousness.[48] Neither is this an exploration of technological weapons[49] or monsters, even though they injure, kill, and create monstrously sick persons.[50]

My ultimate focus—and one of my key arguments—is that technopathologies are not about the technologies themselves but about users. Inspired by the work of Foucauldean feminists studying disease and gender, I take the abnormal as a tool for understanding the normal: how we have constituted ourselves through acts of exclusion.[51] As I will argue in a subsequent chapter, "the user"—including both the term and the people and practices to which it refers—has a long history, one that exists in ongoing processes of entanglement and mutual becoming with related concepts, such as audience. Usership is not a radical disruption brought on with the Internet. However, it is an evolving concept, one in which I argue technopathologies play a key role. Therefore, my use of the term "user" acknowledges its lineage without projecting our contemporary sense of the term back in time. Users and usership, as I approach them, are transhistoric concepts and phenomena, but not transhistoric subjects.

Situational Awareness and the "Psycho Gay Killer" of the Metrolink Train Disaster

The ability to effectively receive, process, and respond to sensory information about our surroundings is known as *situational awareness*, a term derived from military strategy and applied to areas such as transportation, power plant operation, and emergency services. More recently, the breakdown of these processes has begun to be identified as *situational unawareness* or being *situationally unaware*.[52] Often—but not exclusively—this is applied to individuals who are absorbed in mobile devices, its symptoms ranging from antisocial rudeness to putting oneself and others at fatal risk. In 2015, an Amtrak train derailment outside Philadelphia killed eight persons and injured another 200. The National Transportation Safety Board ruled the cause of the accident as the engineer's loss of situational awareness due to technological distraction. He was listening to news about a nearby train struck by rocks and, absorbed in the story, lost track of where his train was, entering a sharp curve at excessive speed. After examining his cellphone records and audio recordings, investigators determined he had not, however, acted with criminal recklessness. Although his technology use made him situationally unaware, it did not go to the extent of making him a criminal.

Almost a decade prior, a similar incident shows an episode in the formation of such an understanding of usership, criminality, and other intersecting social identities. In the case of a California train collision attributed to a texting engineer, usership is clearly not natural, but a constructed and contested terrain, engaging with other social stereotypes and hierarchies. Disease here is one of the vectors through which the user is shaped and understood. His condition of *situational unawareness* shows how pathologization manifests more broadly than simple infection or injury. Furthermore, whether linked by wires, electromagnetic waves, legislation, or economies, bad users pose a threat to other users, and thus we see the explicit demonization of the engineer. In this case, we will see how technopathologies individualize users, distract from social and systemic factors, reinforce existing social stereotypes, hierarchize certain users and usage practices, and result in practices of discipline and control.

On September 12, 2008, a Metrolink commuter train failed to stop at a presumably red signal, colliding with a Union Pacific freight train in Chatsworth, a region of Los Angeles in the San Fernando Valley. The worst train crash in modern California history at the time, it resulted in 25 deaths and

135 injuries, many requiring extended hospitalizations. Preliminary investigation and analyses by Metrolink and transportation researchers noted that a possible contributing factor could have been fatigue from the engineer's schedule—he had worked a split shift of $11^{1}/_{2}$ hours divided by a $3^{1}/_{2}$-hour break, meaning he had experienced only 9 hours away from work in which to rest between shifts. Another issue raised was the possibility of technical malfunctions. The stoplight may have actually shown green in both directions, and backup safety systems had definitely failed to prevent the crash. A later analysis of the Metrolink crash by the California Public Utilities Commission disclosed that, a decade prior, Metrolink and other commuter rail systems had requested and received an exemption from federal requirements to post safety signs reminding engineers of speed restrictions at important points in the track. The requirement had been initiated in response to 1996 passenger train accidents; the 2009 CPUC report concluded that the safety signs should have been installed after all.[53]

However, the element that received the most media coverage was that the engineer, Robert Sanchez, had been sending text messages while driving the train. His technologically weakened attention, it was asserted, caused the accident. Subsequently, many mainstream news stories addressed the dangers of text messaging while driving, including a 2009 episode of the Los Angeles public affairs television program, *SoCal Connected.* Their half-hour episode on the issue included a segment titled "Your Brain on Bluetooth." The story shifted focus from text messaging to mobile communication devices in general, presenting MRI brain scans and scientific research suggesting that they impeded normal judgment and reaction times, even when using a hands-free device. The technology made drivers stupid. Entangling with their perceptual senses, it degenerated their abilities to receive and process sensory information, a form of situational unawareness, although the term was not used at the time.[54]

Although Sanchez' phone was not recovered, one of his recipients showed a message from Sanchez time-stamped one minute before the crash. The National Transportation Safety Board placed the message's dispatch at 22 seconds before the crash.[55] Researchers studying the NTSB data determined that the final message was sent seconds *after* passing the signal, meaning Sanchez most likely was conscious when passing (not unconscious from heart attack or diabetic collapse). Although awake, he was theorized to be suffering from "inattention blindness"—an earlier name for situational unawareness—while composing the message, distracted from seeing the signal. Metrolink already had forbidden texting at the controls but, within a few days, the California Public Utilities Commission passed an emergency temporary ban on all cellular communications by train crew. A week later, California outlawed texting while driving.

Mainstream press coverage evidenced a sensationalistic fixation on texting and the dead engineer's personal life; this was adapted in fringe- and far-right media to support racist and homophobic slander. Initially, typical mainstream news headlines focused on texting, such as "Engineer was Texting Before Deadly Crash," "Texting Eyed in Fatal Train Collision," "LA Metrolink Crash Puts Focus on Dangers of Texting," "Feds Look Into Texting Before Deadly Train Crash," and "Text Messaging May Have Caused Train Wreck That Kills 25 in California!"[56] This became the dominant media frame. Two weeks after the crash, a search of Google News for "Metrolink texting" yielded almost 3,000 results, whereas "Metrolink exhaustion" found only 655 and "Metrolink split shift" only 498.

Although focusing on the technology, such mainstream reports, even immediately after the accident, also were quick to pathologize users. The *Staten Island Advance* equated "Texting + Driving = Lunacy."[57] Some pathologized Sanchez' audience: The *Wall Street Journal* probed "Rail Fans in Deadly Light: Subculture Includes Boy Who Texted Engineer in Crash,"[58] and several stories investigated the "fervent, obscure" and "loco" rail hobbyists, or "foamers," about whom "some wonder if their fervor goes too far."[59] This focus on rail fans—typically teenage males—reinforced pathologizing the engineer, who drove the object of their fervor and with whom they texted. Many reports, such as one on CBS News, specified that Sanchez had texted "teenage rail enthusiasts." A FOX News affiliate headline ran: "Report: Metrolink Engineer Texting With Teen Moments Before Killer Commuter Crash." Here we begin to see how, as Sander Gilman has described, negative stereotypes are often invoked to buttress one another. Why was the age of the rail fans relevant? If Sanchez had been texting middle-aged sports fans, would headlines have shouted about "Texting with Baby Boomers" or "Railfan Retirees"?

Clearly, the stories here were associating and reinforcing the stereotypes of technology misuse and degenerate youth. The frequent association of new communications technologies with unruly teens—from delinquents with transistor radios and antisocial Walkman-wearers to upstart young hackers and underage "sexting" panics—added a sensationalistic thrill to the story. Emphasizing message recipients' ages drew upon and reinforced this pattern of youth being particularly vulnerable to communication media (while also, contradictorily, being seen as the authoritative vanguard of new technology use). It made sense that Sanchez was a bad, sick user because his partners-in-crash were recognizable stereotypes of bad, sick users.

The pattern of delinquent, technopathological youth suggests another stereotype, one quickly called upon to further reinforce pathologization of Sanchez: What was he doing texting teenage *boys*? The implication was that the engineer was a pedophile. Sanchez was gay, and male homosexuality is stereotypically conflated with pedophilia. Conservative talk radio and white supremacist websites picked up mainstream media's investigation of Sanchez' personal life to associate such negative stereotypes and reinforce pathologizations of race, sexuality, ethnicity, and mental health to create a portrait of Sanchez as, not merely a sick user, but a sick *man* deserving of wholehearted condemnation for causing—perhaps intentionally—the accident. All of which pre-empted and defused consideration of other causes of the accident.

Media pathologization of Sanchez began with stories such as the Associated Press' "Metrolink Engineer's Life Held Challenges."[60] Although sympathetic in tone and ostensibly a human-interest tale providing personal background, its content was quite dark:

> In the years leading up to his death in the locomotive of Metrolink 111, engineer Robert Sanchez's life was marked by personal tragedy, jail time and concerns about his health and job security.
>
> His HIV-positive companion had committed suicide, he was concerned about his diabetes, and he feared a brush with the law could end the career he loved.
>
> The National Transportation Safety Board is looking at Sanchez's background after determining that human error and not mechanical failure was likely to blame.

Note the juxtapositions of physical illness (HIV, diabetes, health concerns) with character failings (jail time, job security, suicide) in framing the cause of the accident as human error. Human error could have just as easily been distraction from a text sent *to* him by a pestering, 85-year-old, trainspotting, retired woman, transplanted from Manchester to Chatsworth. However, the background tragedies this article and others used compounded pathological patterns upon Sanchez, and thereby justified suspicious leaps to greater assumptions about him. He was "friendly" but "private and quiet," a "recluse." In 2002, he pleaded guilty to stealing video game consoles. Sanchez had wanted to break up with his HIV-positive partner but, in 2005, his partner had hung himself in their garage—on Valentine's Day, the morning after the couple had argued.

The *Los Angeles Times*' "Metrolink 111 Engineer Led Solitary Life Marred by Tragedy" showed a picture of Sanchez smiling and cuddling a puppy, yet pushed the pathologization further.[61] He was not merely friendly and quiet; he was "relentlessly upbeat" and "intensely private," "polite but guarded," and "different." Although it acknowledged the split shift and diabetic shock as possible contributing factors, these were surrounded by sensationalistic personal details of little relevance other than discrediting the engineer: Sanchez led a "nomadic" life, with recent addresses in six states. He and his partner had bought a house together in 2000, but "No one in [his partner's] family knows how the two men met." Reporters acquired and quoted from his partner's suicide note, which began darkly, "Rob, Happy Valentine's Day." His former lawyer told reporters that, during Sanchez' video-game-theft jail sentence, "'He was going through some personal issues—he didn't tell me what they were. … He said that's the reason he wasn't able to make good judgments. He said a lot of things were going on that caused him to make stupid mistakes.'" Sanchez had received three traffic citations and federal and county tax liens. He had a friend in the greyhound-breeding community whom he took to lunches at "Thai, Brazilian and other ethnic restaurants" but "was reluctant to talk [with her] about much besides dogs and trains."

This last quote was especially disingenuous, as it came immediately after a paragraph in which the friend clearly said that Sanchez often *did* talk about other topics, such as his partner, house, childhood, animals, and 4-H experiences. Yet, the article portrayed as weird his socializing with a woman in her 70s and doing "unusual" things like wearing a tuxedo to a dog show, which any viewer of the Westminster Dog Show will know is not unusual. Another neighbor reported, "Sanchez once told him that he knew some teenagers enamored with trains that he'd occasionally wave to on his route," and the lengthy profile concluded with the specter of Sanchez messaging the "teenage rail enthusiasts."

The stereotype of the lonely, wounded gay man courting the enthusiasm of enamored teenage boys is not subtle. As a nation, we continue to be in the grip of a panic around technologized pedophilia, one that, as Alice Marwick notes, has been disproportionate to actual threats.[62] Certainly, without excusing child abuse in any form, the bogeyman of a sick sexual-predator stranger connecting to victims through communication technologies helps confirm and reinforce the phenomenon of a sick technology user—despite it being statistically rare.

With Sanchez, the biological illnesses, mental illnesses, criminal background, and suspicious details all reinforced and drew the technopathological pattern into recognizability. Race and sexuality were more overtly pathologized as the story expanded beyond mainstream reports to reader comments,[63] websites, blogs, and alternative media. For example, a Boston gay news organization drew on the *LA Times* and AP articles to push the suicide angle with "Bereaved Gay Man at Controls in L. A. Train Crash,"[64] not explaining why Sanchez was still bereaved over five years after his partner's death.

A clearer example of the impact of the *LA Times* story can be seen in the coverage of the crash by KFI AM 640, a talk radio channel broadcasting in the Los Angeles and Orange County areas. According to 2006 Arbitron ratings, KFI was the most popular radio station in Los Angeles, and the most popular AM radio station and news/talk radio station in the country. On September 15, 2008, morning commentator Bill Handel covered the crash with reporter Eric Leonard. Although they mentioned Sanchez, they also noted that the NTSB had criticized Metrolink for prematurely blaming the engineer. The focus of their report was on how the NTSB had been arguing for 30 years to install Positive Train Control, a safety device used in some Northeastern routes that shuts down trains if an engineer runs a red signal light. According to the NTSB, it would have prevented the Metrolink train from subsequently breaking through a rail switch and colliding, but railroad and transportation companies have resisted the expense of installing the system.

On September 17th, however, the *LA Times* article appeared. KFI's self-described "conservative political commentators" John Cobell and Ken Chiampou covered the case with the same reporter, Eric Leonard, on *The John and Ken Show*. Now the story was all about Sanchez, his "strange life" that included "a gay lover" and was "so creepy." They discussed the *LA Times* article, giving out names of all involved, quoting the most lurid details, even reading the suicide note twice. They noted that Sanchez had been at the throttle of a train recently that had killed a suicidal pedestrian who had walked in front of it. They discussed how NTSB evidence suggested Sanchez had run the red signal light, broken the switch, and then slowed the train down to go into a tunnel, but had not applied the brakes in final seconds of crash. This suggested he was conscious and therefore intentionally caused the crash. Another pathology: suicidal.

Leonard added that, by "bizarre coincidence," he personally knew the relatives of Sanchez' lover who had killed himself, the Burtons. A wealthy Beverly Hills family, Leonard reported that the family believed their son was murdered by Sanchez for having given Sanchez HIV, and cited evidence they uncovered through a private investigator. The family had called Metrolink to report that Sanchez was unstable, and presented their evidence to police, but both the coroner and law enforcement had ruled it a suicide.

"He could be capable of pretty horrendous things," the hosts mused, "a psycho case." Chiampou said,

> People can put on all kinds of façades. People do that every day. … You combine that with him running over somebody who, I guess was trying to commit suicide by walking in front of a train a few weeks back. … I mean there's a lot of darkness and ugliness, and maybe this isn't the end of the story. Who knows what we're going to find out.

The hosts took a call from Craig Williams, the father of a 14-year old boy who had been texting Sanchez before crash, and who had been interviewed and covered widely in the news media. The first question the hosts asked Williams was, "What was your son's relationship with Robert Sanchez?" The father spoke of the rail fan community, but one host pressed the issue: "You know a lot of people, when they hear about a 46-year old man, who lives alone, not married, he's texting their teenage boys-" Williams, a fan of the show, then off cut off the host. He claimed he wanted "to set the record straight" that the cause of the wreck was *not* texting but a stroke or diabetic coma. The hosts disagreed, describing Sanchez as "obsessed with texting," and suggested he didn't brake at impact because he was texting. Williams explained that he wanted to "shift directions" of the story away from Sanchez' personal life. He blamed a TV reporter for misrepresenting his son's interview and turning it into a story about texting, and he went on to describe how his son has been receiving "nasty responses" on his train website and had been forced to shut it down. People had been blaming him for the crash, and he had been forced to miss school for a day due to the negative attention.

A conservative blog sourced the KFI talk radio reports in a post titled "KFI: Metrolink Crash Engineer was Psycho Gay Killer,"[65] amplifying, not only the suicide angle, but also the family's claim that Sanchez had murdered his partner. Conservative political group (and supporter of a campaign against California gay marriage at the time), Campaign for Children and Families, bannered their website with "LA Train Engineer was an Unstable Homosexual."[66] A website associated with white supremacist David Duke, *WhiteCivilRights.com*, ran the headline "Gay Latino Train Engineer Texting Boys."

As the investigation progressed, reports emerged that Sanchez had once put a teenage boy at train controls, an illegal and unsafe practice, but, moreover, one which called to mind the sodomy-like pose of Sanchez standing behind the boy, perhaps even with arms around him or hands on his shoulder.[67] Another white supremacist site, *NewsNet14* ("Global and Local News for Europeans Everywhere") ran with "Gay Hispanic Train Engineer Crashes Train while Texting Young Boys, Kills 25."[68] One conservative blogger quoted the entire *CBSNews.com* article, "Details In Train Engineer's Troubled Past Emerge," as support for his own post titled, "Homosexuality is a Dangerous Mental Disorder. Study: Homosexual Lifestyle Strongly Linked to Depression, Suicide," using Sanchez to reinforce and revive archaic diagnoses of homosexuality as mental illness. This demonstrates how, despite medical authorities dismissing a diagnosis, it still has cultural relevance.[69]

Evidence emerged that continued to suggest the signal light may have been malfunctioning, and it also came to light that the driver of the other train tested positive for marijuana. To appreciate the misdirected pathologization in the case of Sanchez, consider the case in the abstract: There is a vehicle crash. Of the two drivers, one tests positive for substance use. Which driver, then, regardless of personality characteristics or private life, would one expect to be the focus of suspicion, investigation, and media coverage? Clearly, the drug user. Yet, when this detail emerged, typical media reports were similar to a March 3 local network evening news broadcast, which, in reporting on the ongoing NTSB hearings, spent nearly the entire report on Sanchez, not mentioning the other, pot-smoking engineer until an offhand comment in the third-to-last sentence of the report.[70]

Sanchez did not test positive for drug use, but, revealingly, members of one of my focus groups transferred this detail to him (in addition to nearly tripling the death count and mistakenly imagining the other conductor as also texting). At first, they had been talking negatively about text messaging in general, and expressed concern for how texting had distracted Sanchez—blaming the technology. This switched when the pedophilia allegation was brought up, and was then reinforced by the incorrect association with drug use—blaming the user. The chain of pathologizations helped shift blame from systemic issues and the technology to an individual:

FEMALE:	Wasn't there a story about a train that—and the texting.
FEMALE:	Yeah.
	. . .
MALE:	Yeah, the whole [inaudible]

...

FEMALE: I don't know the total details, but I just knew that they thought it was caused because he was texting—(agreeing at once)

FEMALE: He sent one 15 seconds before the—

MALE: He didn't hit the switch. The switch would have put the trains on different tracks, but he was texting and passed the switch out and bam.

MODERATOR: Did anyone not hear of this story?

FEMALE: I didn't.

MALE: How many people died? I don't remember how many people [inaudible]

MOD: Quite a few of them.

MALE: 65 people died.

MALE: Yeah. It was pretty bad.

FEMALE: Too many.

MOD: Did anyone hear in the news coverage on this, or in talking to other people, of any other factors that were involved in the accident, other than the texting?

MALE: I thought they said somebody was in there with him.

MOD: Somebody was in the—

MALE: Yeah.

MALE: That he's had someone in there before.

MALE: Oh, yeah, that's what I mean, before.

FEMALE: Yes.

MALE: [inaudible] that was on a train, and he actually took on a ride. (agreeing at once)

MALE: And that supervisor got fired over that.

MALE: But there was another guy-

FEMALE: The text message-

MALE: There was another guy in another way going the opposite way who was texting him as well, texting him back.

MOD: In the oncoming train?

MALE: I think it was the oncoming train, and you know, so they were both texting each other, and they have per second the last text in between.

...

MOD: Did anyone hear anything about anyone involved in that accident?

MALE: Yeah, he tested positive, didn't he?

MOD: I'm asking-

MALE: I heard he tested positive. I work for [governmental agency].

FEMALE: Positive for what?

MALE: Marijuana.

FEMALE: Oh.

MALE: That's what I heard.

MOD: Who was it that tested positive?
MALE: The driver of the train that was texting. That's what I heard. I don't know if it's true or not.
MOD: Anyone else heard? (talking over)
MALE: [A person I know] died on that train.
MOD: Oh, I'm sorry to hear that.
MALE: And he tried talking her into going for the train ride that night, and she had to stay at work and work overtime.
MOD: Did anyone hear anything else about this engineer and the accident?
MALE: I didn't hear anything about substance abuse.
FEMALE: Neither did I. They might have kept that out of it. [the media]
MOD: Anything else about him or his personal life or anything?
FEMALE: Well, I heard that he was depressed. You know, he had a rough life, going through some depression.
MALE: He was texting a 12 year old.
FEMALE: Yeah, he was texting young kids.
MALE: A lot of minors, young kids.
FEMALE: Really? Because I didn't hear that.
FEMALE: Yeah, young boys which is-
MOD: How many people had heard that aspect? [counting] Two? [others saying no at once]
MALE: Which one, the substance thing?
MOD: Well no, that the people he was texting when he was-
FEMALE: About texting young kids.
MALE: So he was a pedophile? Yeah, I didn't hear that.
FEMALE: Well, you don't know. He may have just gotten along with them.
MALE: If you're texting little kids, it's a problem.
FEMALE: He was smoking drugs.
MALE: When they're not your children.

Pathologization works to obscure systemic factors: the transportation infrastructure, the working hours and conditions, the mental health of a gay Latino widower in a racist homophobic culture with no universal mental healthcare. Even if, for example, homosexuality was scientifically correlated with depression and suicide, it would not mean that homosexuals were inherently sick to any greater degree than the conditions that a homophobic culture that could understandably drive them to depression and suicide. Or, for another example, one comment to the *LA Times* article[71] about Sanchez protested:

> the nomadic lifestyle? hey, come'on. this is what happens when every job out there is "temporary" and the idea of a real working class "career" with the same company in the same location is just kaput. This is how people live nowdays.

The train engineer is a paradigmatic figure of America, the West, and all its national mythos, as well as masculinity, according to Lynn Kirby in her study of railroads in early silent cinema.[72] If so, then a gay Latino engineer in California, a Western state embroiled in a gay marriage debate, and, more precisely, Los Angeles, home to many immigration protests and controversies at the time, was in many ways a challenging image. In Los Angeles, trains historically have been a tool for race- and class-based segregation, with public subway lines stunted, stopped, and rerouted to prevent easy access to white, upper-class neighborhoods. Sanchez offered a scapegoat for anxieties around these changing social roles and ongoing debates. The rush to pathologize and condemn him evidences longing for traditional notions of nation, masculinity, and sexuality. Moreover, there is a long history of using ethnic minorities as negative examples when teaching proper usership. "Sanchez at the Throttle" historically echoes "Cohen at the Telephone," the first-ever million-selling comedy record in the US, which was a 1916 sketch of a stereotypical Jewish man's poor usage of the then-new device.

Pathologization individualizes in a way that obscures various social factors involved. The sick user obscures the lack of social support and mental health care for a grieving gay man, which perhaps encouraged him to seek out friendships and admiration from texting railway fans. Working split shifts, he may have texted or socially communicated in other ways on the job because there was little other appropriate time available. Depressed and lacking adequate mental healthcare, he may have craved affirming human contact.

The ideals of mediated communication as human connection show a dark side when connection becomes contagion. Pathologization marks Sanchez as a degenerate—an archaic medical condition I will discuss in a later chapter. Key here is that, as Foucault describes, "The degenerate is someone who is a danger. ... The degenerate is someone who, at all events, cannot be cured."[73] The implied threat is of Sanchez infecting the boys through the homosexual predator stereotype. Sanchez is positioned as a degenerate, not a victim of an evil texting technology. Sanchez the "loner" and outsider has tinges of social withdrawal in his narrative and, in networking with inappropriate or at least atypical associations with teenage boys and older women, raises the specter of new social reconfigurations.

Moreover, the case also disputes the newness hypothesis. Texting antecedents can be traced back to experiments with pagers in 1989. The first use of a cell phone network is attributed to 1992. U.S. cellular carriers began allowing interoperability of messages between networks in 2001–2002. By 2007, texting was the most widely used mobile data service, with 74% of all mobile phone users worldwide being active users. An article in *Wireless Week*, published less than four months after the Metrolink disaster, called it the "Year of SMS, Again." It noted that, "For the past few years, SMS traffic in the United States has been growing at a rate of over 125% annually. Sixty-six percent of mobile phone owners use SMS. … In 2008, the number of messages sent surpassed the number of voice calls made."[74]

Therefore, what Sanchez was doing was perfectly normal. We are constantly bombarded with incitements to do these exact things. In my former advertising career, I wrote copy and devised brand strategy for a major wireless phone company in their marketing campaigns celebrating always-on, mobile connectivity. They and others had been promoting texting and mobile data to pursue the revenue streams first seen in Asian countries, replacing profit lost from disconnected landlines and increasingly generous voice-minute packages. Academics have celebrated the people-power of texting "smart mobs" to challenge global banking oligopolies and topple governments,[75] while popular culture has wallowed in the joy of text-synchronized flash mobs, dancing wildly in public places. Given such exhortations for technology use, how can anyone blame the victim for doing exactly what they were told to do, in so many circumstances, from so many corners? Technopathologies are dissimulation from not only systemic factors involved in accidents, but from the very choruses of exhortation to use the media and communication technologies that they exemplify. The pleasure of multitasking, the feeling of facility and mastery that comes from being doubly productive—whether in job tasks, personal tasks or, perhaps even more likely, efficiently combining the two—is reflective of the individualism and flexible labor of neoliberal social discourses. We are expected, if not continually urged, to blur the boundaries between work time and personal time, flexibly combine the two, and find pleasure in heightened efficiency and productivity.

Sanchez also clearly shows the normalization of disease as part of technology use. Texting distraction—a phenomenon not limited to driving, but one that has been applied to pedestrians as well—is not so different from ADHD. What is distraction, after all, but faulty attention? For, if not exactly a deficit, it is a deficit of attention directed to the things deemed proper. Situational unawareness is attention misdirected—a tremendously important mismanagement of resources in today's media-saturated, networked "attention economy." And, as the rapid and widespread passage of texting laws indicates, this has rapidly become a normal part of usage practices. At the time of writing, 46 U.S. states, the District of Columbia, Puerto Rico, Guam, and the

U.S. Virgin Islands now either have bans or restrictions on using handheld devices, texting, or accessing the internet while driving. It is no longer that some people—youth, minorities—are degenerates or weak, dangerously susceptible to texting's seductive siren. All users are, and this determines the laws dictating what all users must do.

As I will discuss in a subsequent chapter, the historic and ongoing associations of sickness and femininity, from neurasthenia to today, can be seen in the prurient interest in Sanchez' sexuality, love life, and portrait as a grieving widower, suggesting the feminization of the user. If all users are femininely unstable, and sick or potentially so, then all users warrant regulation and control. Which, in the case of distracting driving, is exactly what we have seen occur. The texting-caused car accident is now so familiar it appears in public service announcements and popular films such as 2010's *Limitless*. A *Saturday Night Live* television commercial parody promoted "Headz Up," a smartphone app that texts pedestrians real-time warnings when they are about to step into traffic. A report of it on a financial website emphasized the user pathologization: "Feel free to ignore the real world right in front of your face and go about the important business of texting, Twittering, searching, updating your Facebook status, or playing Words with Friends or Farmville, Headz Up will be there to let you know when you're in danger or there's something right in front of your face that you should be aware of." A 2010 study attributed 28% of crashes to not only texting but cellphones in general, and a report at the same time reiterated that hands-free interfaces and attachments provided no significant reduction in distraction.[76] *Business Week* reported that, by 2012, "The number of headphone-wearing pedestrians seriously injured or killed near roadways and railways has tripled in six years."[77] By 2013 discussions had begun about whether wearing computing devices, such as Google Glass, presented a threatening new form of distraction.[78] A 2014 study found the "cognitive distraction" of hands-free, voice-activated applications and car control systems increased car crash risks more than handheld devices and hand-operated systems.[79] By 2016, the mobile game Pokémon Go was blamed for over 110,000 road accidents in the U.S. in only 10 days,[80] and a lawsuit was filed over a crash caused by a driver using Snapchat's speed filter.[81]

Tracking Technopathologies

Although I am tracing the development of discourses, not devices, my methodological approach is amenable to Claude S. Fischer's "user heuristic," an anti-determinist perspective in social historiographies that examines technologies from the perspective of user experiences.[82] My project is about users and their experiences of ill health, as communicated across a variety of primarily popular, public discourses. With this perspective, I employ a *queer methodology*. Initially this was described as an interdisciplinary combination of "textual criticism, ethnography, historical survey, archival research, and the production of taxonomies," which rejected the social science method of trying to "squeeze truth from raw data."[83] More recent scholarship has clarified the epistemological and ontological underpinnings. "If, as queer thinking argues, subjects and subjectivities are fluid, unstable and perpetually becoming, how can we gather 'data' from those tenuous and fleeting subjects using the standard methods of data collection such as interviews or questionnaires?" Such fundamental destabilization must also involve a reflexive destabilization of the researcher's position as well, which I will address later.[84]

Such projects have a long history in science and technology studies, particularly feminist approaches and critiques. In this book, I often rely on the agential realism of Karen Barad, which combines poststructural, feminist, and queer perspectives with her own background as a trained physicist and historian of science. In the "ethico-onto-epistemology" of agential realism, objectivity is reconceived as, not the researcher's distance, but the researcher's ethical obligation to acknowledge that they are within the world of phenomena they study, not outside of it. Moreover, all phenomena are ongoing processes of becoming, endless iterations existing in mutual entanglement, and the researcher makes arbitrary cuts of classification, exclusion, and definition in order to study particular phenomena and processes. However, these "agential cuts" are artificial, and the researcher must acknowledge this as part of their own process.

If everything is a phenomenon, not a thing, and also entangled with other phenomena, producing new patterns and agencies, how do we see or hear anything meaningful? Isn't it all simply too much to take in? This is where agential cuts are involved. A cut is similar to framing, culling, bracketing, or editing, but with a crucial distinction: it acknowledges this as a process of boundary-making, and not a true separation. It takes responsibility for the process of bringing patterns into legibility, but without resuming the patterns are disconnected from all the other phenomena and material-discursive practices that make up the universe, including apparatuses making the cut.

> Apparatuses enact agential cuts that produce determinate boundaries and properties of "entities" within phenomena, where "phenomena" are the ontological

> inseparability of agentially-intra-acting components. That is, agential cuts are at once ontic and semantic. It is only through specific agential intra-actions that the boundaries and properties of "components" of phenomena become determinate and that particular articulations become meaningful. … Apparatuses are boundary-making practices. (p. 148)

This book is an agential cut. I am making certain boundaries in calling into legibility pathological technoculture and the phenomena of techno-pathologies. But I do not presume that my phenomena contain themselves neatly within the boundaries I have marked. I have marked boundaries for reasons of intelligibility and coherence, but they are not absolute ontological limits. I focus on sickness associated with electric communication technologies, but acknowledge there are non-electrified technopathologies. There are also pathologies of transportation and other non-communication technologies. I focus on cases in the United States, but do not argue that they are unique to the U.S. and, more importantly, acknowledge that "the US" is a permeable boundary, looking at, for example, Japanese novels and foreign films and readily available here. Finally, I do not place myself outside the phenomena. I am part of the apparatus and, through my agential cuts, part of the production of these phenomena.

Pathological technoculture extends across multiple and varied dimensions of social life. Therefore, this book employs a transdisciplinary approach to studying it. I employ cultural studies, particularly as it has become inflected in the field of Communication. Although I draw on a variety of theoretical resources, such as media studies, feminism, poststructuralism, and sound studies, my primary arguments are about social phenomena, not theory. Similarly, although I employ textual analysis, I skew towards an empirical orientation that requires buttressing my interpretations of texts with historical evidence and direct engagement with persons.

Therefore, this project uses a triangulation approach to method, merging together focus groups, critical discourse analysis, and historiography to examine my topic. These are integrated throughout the book, although certain chapters may utilize a distinct version or approach to organizing the topic, such as a single case study, a variety of cases of a specific type of disease, or cases related to a specific type of technology.

My data collection has consisted of building an archive of cases of techno-pathological discourse, now numbering over three hundred examples. I have followed pathological technoculture as it extends across factual, fictional, and indeterminate sources. I have drawn on examples scattered throughout secondary scholarship, as well as numerous examples from my own research in media representations, archives, and interpersonal interactions. I have examined newspapers and scholarly publications, but also popular culture, such as movies and supermarket tabloids. My research design was intended to capture the scope of technopathological discourse—to demonstrate its wide and varied connections. Limiting my search to medical journals or television shows would have suggested that the phenomena of pathological technoculture were limited or uniquely connected to those fields. As electric communication technologies permeate so many aspects of social life, I wanted to explicitly show that their diseases did as well. Furthermore, I was interested in elevating certain neglected and nonauthoritative media in studies of popular culture. Supermarket tabloids, for example, are massively produced and distributed, yet are not readily held in any accessible, comprehensive library or archive, making their use in historical cultural studies particularly difficult.

My research took place in traditional and online archives, general web searching, numerous examples forwarded to me by friends and colleagues, interpersonal discussions online and off, preexisting scholarship, material artifacts (such as instruction booklets and warning labels on packaging), broadcast and cable television programming, VHS tapes, DVD recordings, streaming film libraries, movie theaters, and in focus groups. I conducted a trial group with six participants discussing technopathologies, from which I revised my questions and examples for two more focus groups of ten people each. These had a mix of ages and genders, but all were held in Southern California with participants that were disproportionately Caucasian. Participants were screened for a mix of people who had and had not experienced technopathologies themselves, but who were all regular users of electric communication technologies.

Finally, this queer, agential approach to research design extends to its presentation. I am interested in how the discourses of technopathologies are fluid and unstable, traversing various institutions and sites of knowledge, historiography, and culture. Therefore, I use a method of combination and juxtaposition to present my findings. As in the example of Robert Sanchez, I combine different methods as they serve to provide insight into and support for specific points. To isolate textual analysis, historiography, or focus groups into separate chapters, as is often the custom in academic publishing, reinforces a false sense of stability—that one method is used to reveal one argument or dimension of it. It isolates research as discrete acts, replication and accumulation notwithstanding. I wish to employ an agential perspective instead that foregrounds the entanglements and connections between phenomena and processes I am investigating. To create false separations is, moreover, simply inaccurate, denying the extent to which these discourses travel over social life. This project is focused on users, and their messy, complicated social lives. My methods are integrated throughout this book, rather than chapter by chapter, to reflect this social experience.

This also extends to my agential cut of organizing the project by evolving dimensions of usership, instead of my specific technologies. To focus on individual technologies ("Sicknesses of television," "Diseases of the internet") would enact the technological determinism this project avoids. I am not telling a story about technologies; I am telling a story about their users. I want to listen to them and counter existing technological narratives, such as the newness hypothesis, by demonstrating how technopathological discourses are about users, not scary new technologies. Also, I am attempting to stay true to the evidence of my artifacts. Although there are certain broad generalizations one can make, particularly in regards to mass media technologies versus more interactive, individualistic ones, technopathologies cannot be isolated and grouped by device. Radio, TV, text messages, wired phones, mobile phones, videotapes, and computers all carried toxic curses and supernatural demons that possess, hurt, or kill sufferers. Both the telegraph and the original iPod scroll wheel caused thumb injuries. iPods increased susceptibility to lightning strikes, as did telephone wires and selfie sticks. Movie theaters and telephone receivers transmitted germs, as selfie-taking spread head lice; movies and television both encouraged delinquency. Other groupings, such as analog versus digital, are difficult to maintain as well. For example, portable analog cassette players and portable digital mp3/AAC players both caused distraction, vulnerability, hearing loss, and anti-sociality.

Overview of Book

Bearing these methodological commitments in mind, the book is laid out in the following manner. The first chapter will present how pathologization is a technique of constituting subjects. Using the historic proto-technopathology of neurasthenia, I will show how sickness performs cultural work of categorizing along the lines of gender and abnormality. In Chapter 2, I turn to the specific subject category of users. I clarify my use of the concept of users, arguing that it is tied to interactivity in communication technologies dating back to the telegraph and telephone. This is in contrast to, but simultaneous with, mass-media audiences. Rather than a disruption brought on by computerization, or an oppositional dichotomy, users have long coexisted with audiences, and are better understood as a neoliberal acceleration, rather than disjuncture. Concepts such as interactivity and "Web 2.0" are historicized and critiqued. Using the example of telephone training films, I demonstrate how users are instructed on proper usership and bad users are pathologized through connection to existing social stereotypes.

In Chapter 3, my argument progresses to show how technopathologies are far from ridiculous or irrational but draw on long-standing traditions in science, medicine, and religion. The chapter presents an examination of the cultural roots of technopathologies with a historical discussion of the element joining the technological and the biological: electricity. Drawing on primary and secondary sources, I examine electricity's role in health and illness across sites such as spiritualism, electrotherapy, physics, and medical science. Chapter 4 extends this complication of the newness hypothesis with an extended discussion of the town of Anderson, South Carolina, "The Electric City." Anderson was a regional and national leader in early electrification and illumination yet, instead of the utopian celebrations or dystopian panics that histories of electrification and technology emergence would lead one to expect, examination of historical documents reveal decided ambivalence. In what I call "the electrical banal," we see a portrait of relatively rational reactions to technology, even when beliefs in electrical relations to spiritualism, the supernatural, and health are present. Yet technopathologies—what one would expect if they were simply irrational reactions to new technologies and all they represent—are largely absent.

Having established that technopathologies are not necessarily irrational, and that new technologies do not necessarily provoke irrational reactions, Chapter 5 presents evidence of the historical continuity of technopathologies. Here, I demonstrate that technopathologies are not merely reactions to new media. I will present several cases of sickness associated with dominant, domesticated technologies and even old media. My argument expands on a second front as I argue that this indicates the individualization effect of technopathologies. Pathological technoculture does not demonize technologies, but users. Using the contemporary example of electrosensitivity, I show how technopathologies are morality tales of bad users and thereby constitutive discourses of what it means to be a user.

In Chapter 6, technopathologies become social diseases, but still individualized with a eugenic logic. Sick users are not victims but threats who could infect good, healthy users, and also reproduce more bad users. This is described using the television consulting and advocacy of actor Robert Montgomery and a cluster of media texts derived from Koji Suzuki's techno-horror novel *Ring*. In Chapter 7, I employ Priscilla Wald's work on outbreak narratives and contagion to examine technopathologies in which users withdraw from society into technologies. Focusing on mobile audio technologies, I examine the case of the Walkman-wearing "Central Park Jogger" as a carrier of feminism, and explore a variety of associations between personal audio and "degenerate" populations. Taking these cases to their logical extreme, in a variety of fictional narratives in which technologies turn users into sociopathic monsters, I argue that the real threat of sick, withdrawn users is not their social disengagement, but their potential reorganization into new social formations.

In the conclusion, I argue that the abnormality of disease has become normalized in the sense of being a common and defining aspect of usership. Paradoxically, sickness is a typical state, requiring constant surveillance and control, although the responsibility for this varies. Through an extended close reading of four ergonomics diagrams designed to prevent repetitive stress injuries at computer workstations, I argue that the evolving position of the user is a feminized subject with ongoing self-monitoring and adjustment along multiple dimensions, because she is presumed to be already sick.

Notes

1. E.g., Glassner, *Culture*.
2. Taussig, "Reification."
3. Barad, *Meeting*. See also Bashford and Hooker, "Introduction," Hutcheon and Hutcheon, *Opera*; Pernick, "Contagion Culture," Rosengarten, "A Pig's Tale."
4. Lupton, *Medicine*.
5. Akrich and Pasveer, "Multiplying Obstetrics".
6. Conrad and Schneider, *Deviance and Medicalization*.
7. Chouliaraki, *The Spectatorship of Suffering*.
8. Foucault, "Dangerous Individual."
9. E.g., Balsamo, *Technologies*, Banet-Weiser and Portwood-Stacer, "'I Just Want'," Cartwright, *Screening the Body*, Hartouni, *Cultural Conceptions*, Hayles, *My Mother*, Serlin, *Replaceable You*, Van Dijck, *Transparent Body*, Weber, "Makeover as Takeover," "What Makes the Man?".
10. De la Peña, *Body Electric*.
11. Leslie, "Google Makes Us All Dumber".
12. Linshi, "iPhone Separation".
13. Firger, "Depression".

14. Gallagher, "E-books 'Damage Sleep and Health'".
15. Rettner, "Man Tears Tendon".
16. Subramania, "Casual Sex Apps Blamed".
17. Ackerman.
18. Associated Press, "Temporary Blindness Tied".
19. Hernandez.
20. Agency, "Mobile Phones Are Cooking Men's Sperm."
21. Pascual, "What Is 'Phubbing'?".
22. Salinger.
23. Allon, "Is Your Smartphone Aging Your Face Prematurely?".
24. Ibid.
25. Ibid.
26. Brustein.
27. Oswald, "Upgrade Effect".
28. Brody, "Screen Addiction".
29. Contrera, "The Secret Lives".
30. Estes, "My Smartphone".
31. Timberg, "How Video Games Suck You In".
32. Lee, tech2 News Staff.
33. Kardaras, "The Dangers of Pokémon Go".
34. AXXIDENTAL, "How Is Your Phone Changing You?".
35. Totten, "Kill Shot".
36. Horton, "More People Have Died".
37. Zargar, "Major City".
38. Allon, "Is Your Smartphone Aging Your Face Prematurely?".
39. Skultans, "Historical Disorder."
40. Allon, "Is Your Smartphone Aging Your Face Prematurely?", Crist, "Leaning Forward During Phone".
41. Boorse, cited in Parsons, *Social System*; Turner, *Body*, 176.
42. Lutz, *American Nervousness*.
43. CBA/AP, "'Exorcism' in Morgan Freeman".
44. Dunn, "The Power of Christ".
45. Kennedy, *History*, OED, "Disease."
46. Cvetkovich, *A Public Feeling*.
47. Pelling, "Meaning of Contagion."
48. Unlike, for example, Kenneth Burke's "technology psychosis" in *Permanence and Change*, 38–42, Heidegger's famous "essence" of technology ("Question") or Feenberg's attempt to operationalize this (*Questioning*).
49. Even the most direct disease relationship, that of causation, does not necessarily imply intent. In fact, I suggest that one of the defining aspects of disease is the lack of conscious intent. This is what differentiates disease from weapons. Smallpox variola virus and arsenic do not intend to hurt human beings. The intent of microorganisms is merely to survive and replicate, with damage to human hosts a mere consequence. They are part of the natural world that inherently cause harm to the human body. A gun, however, is unnatural:

something created with the intent of realizing human intention to cause harm to another living thing or object. Of course, agents of sickness can be weaponized. Someone could poison you with arsenic or intentionally expose you to smallpox. But weapons always require intent. Sickness does not. In fact, this is part of their horror: they injure, sicken, and kill for no reason other than it is their elemental nature. Their very being is deleterious, and you can encounter them through no action or provocation of your own. Disease can be weaponized, but weapons are not diseases. Devices for communication transformed into tools for intentional death and injury can be seen in numerous films: a death ray in *Murder by Television*, a cell-phone bomb detonator in *The Dark Knight*, bludgeoning by TV set in *Halloween* (2007) or computer keyboard in *Gattaca*, strangulation by phone cord in *Halloween* (1978) or Walkman headphone cords (twice) in *The Living Daylights*, cerebral impalement from a spike-emitting phone receiver in *Dr. Phibes Rises Again!*, *Mikey*'s boom-box-into-hot-tub electrocution, the death contraptions triggered by the number of visitors to a snuff website in *Untraceable*, and impalement by an ocular telephoto video lens and razor-sharp flying discs from a chest CD-player of the supernatural industrial cyborgs in *Hellraiser III: Hell on Earth*. Comic variants include the diePod euthanasia machine of *The Simpsons*' "Million Dollar Abie" episode or the alien TriPod of *Scary Movie 4*, with its "Destroy Humanity" playlist. Thinking of technologies as sickening helps maintain a conception of technologies as social assemblage and thereby not neutral. In contrast, the conceptual framework of weaponry would conceive of technologies as neutral vessels for human misuse or malfeasance. ("Guns don't kill people, people kill people.") True, technologies can carry out the politics of their creators, as noted in concepts such as Langdon Winner's artifactual politics or the technological affordances of actor-network theory. Disease is different in that the effects are not the intent of the designers or inventors. They are the device's unintended consequences or uses. Disease also is symbiotic; it requires the sickening agent *and* the susceptible patient. Both are implicated. I attend to the user, and usage is by definition relational. Disease also maintains a bodily focus, which is important for several reasons. For one, this is intended as a corrective to utopian technological discourses of bodily transcendence, which dangerously elide corporeal issues of labor, gender, and pain. In addition, this maintains a bodily focus. In technopathological discourses, the user's body is a site of contestation; the diseased user is anything but disembodied.

50. Diseases and their carriers may be horrifying and monstrous, but they are not monsters. A monster injures and kills, but injury and death do not have the same ontology as sickness. Injury and death are actions, things that happen, whereas sickness is a state of being. Clearly, they can be related, but for my understanding of pathological technoculture, they should not be conflated. The great white shark in the movie *Jaws* is a living creature that is a monster who injures and kills people, but "shark bite" is not a disease. The microorganism that kills millions of people in the movie *Contagion* lacks the personality or identity of a monster, although it certainly is monstrous. It does not kill so much as create a state of sickness that leads to death. However, a monster can be the result of a disease: the "rage virus" of *28 Days Later* infects people and turns them into zombie-monsters. A mental illness such as sociopathy may express itself in the actions of a serial killer, but the monster is the killer, and the illness is the killer's disease-state. Thus, for this project, I differentiate between a pathological technology that induces, enables, or exacerbates a sickened state

and a technology that has become a sentient, attacking monster. Therefore, I exclude examples of technological monsters, such as the killer robots or computers in films such as *2001: A Space Odyssey*, *Colossus: The Forbin Project*, *Saturn 3*, *Virus*, or *Gog*. These categories of disease, monster, and weapon are, of course, more heuristics that absolute divisions.

51. Bordo, *Unbearable*; Bartky, "Foucault, Femininity"; Foucault, "Political Technology," "Subject and Power"; Lasch, *Narcissism*.
52. Schrey, "How to be Situationally Unaware," Sellers, "Are You Situationally Aware?"
53. Connell, "Metrolink," "Governors Highway Safety Association."
54. This is less unusual than it may sound: Various forms of attention deficit and concentration problems are recognized as diseases or symptoms. Many people have blamed this on the programming and editing styles of contemporary movies and television, which have encouraged shortened attention spans. This, then, fits the familiar pattern of a disease caused or worsened by technology. It is also a similar pattern to the fatally divided attention and concentration of engineers, drivers, and even pedestrians who are texting.
55. The NTSB had just concluded investigation of a May 28, 2008 rail crash in Newton, Massachusetts, determining that similar rumors about engineer-texting were unfounded.
56. Associated Press, "Engineer was Texting"; Sahagun and Rohrlich, "Texting Eyed"; Wood, "LA Metrolink"; Associated Press, "Feds Look"; Dij, "Text Messaging."
57. "Texting + Driving = Lunacy."
58. Audi, "Rail Fans."
59. Gold, "Crash Shines," "Train Hobbyists"; Hennessy-Fisk and Gold, "'Railfans'."
60. Risling, "Metrolink."
61. Quinones and Kelley, "Metrolink."
62. Marwick, "MySpace."
63. Most of the 90 comments posted to the LA Times article in the 48 hours after it was published were critical of this coverage: "los angeles times is really empty of a soul." However, a few did speculate that Sanchez' diabetes or possible act of suicide made him and Metrolink responsible.
64. Melloy, "Bereaved Gay."
65. Varones, "KFI: Metrolink."
66. "Off the Rails."
67. The outrage over a visitor in the control room seems less justified when one recalls that inviting children into the cockpit of airplanes used to be common practice, famously satirized in *Airplane!* with pedophilic humor.
68. "Gay Hispanic Train Engineer."
69. Mathis, "Homosexuality."
70. KCBS, *News*.
71. Quinones and Kelley, "Metrolink."
72. Kirby, *Parallel Tracks*.
73. Foucault, "19 March 1976", 317–318.
74. Shah, "2009: The Year."
75. Rafael, "Messianic"; Rheingold, *Smart Mobs*.
76. Halsey III, "Hands-free."
77. Kelley, "Study: You are More Likely"

78. Streitfeld, "Wearable."
79. Halsey III, "Hands-free."
80. Borland, "Don't Pokemon Go and Drive!".
81. AP.
82. Fischer, *Social*, 17.
83. Halberstam, *Skin*, 10.
84. Browne and Nash, *Queer Methods and Methodologies*, 1; Kulpa and Liinason, "Queer Studies", 2009; Liinason and Kulpa, "Queer Studies", 2008; Lööv, "Queering Methodology", 2012.

References

2001: A Space Odyssey. Dir. Stanley Kubrick. 1968, UK/USA.

28 Days Later. Dir. Danny Boyle. 2002, UK.

Ackerman, Dan. "Touchscreen use linked to sleep problems in infants." *CNET*, 2017, April 13. https://www.cnet.com/news/new-study-touchscreen-use-linked-to-sleep-problems-in-infants/. Accessed November 21, 2017.

Agency. "Mobile Phones are Cooking Men's Sperm." *The Telegraph*, 2016, Feb. 22. http://www.telegraph.co.uk/news/health/news/12167957/Mobile-phones-are-cooking-mens-sperm.html. Accessed July 23, 2017.

Akrich, Madeleine, and Bernike Pasveer. "Multiplying Obstetrics: Techniques of Surveillance and Forms of Coordination." *Theoretical Medicine and Bioethics*, vol. 21, no. 1, 2000, pp. 63–83.

Allon, Janet. "Is Your Smartphone Aging Your Face Prematurely? 5 Maladies of the Digital Era." *AlterNet*, 2015, August 24. http://www.alternet.org/personal-health/your-smartphone-aging-your-face-prematurely-5-maladies-digital-era. Accessed July 23, 2017.

Associated Press. "Engineer was Texting Before Deadly Crash." *San Jose Mercury News*, 2008, September 18. http://nl.newsbank.com/nl-search/we/Archives?p_action=listandp_topdoc=21. Accessed May 29, 2009.

Associated Press. "Feds Look into Texting before Deadly Train Crash." *Sulekha.com news*, 2008, September 17. http://newshopper.sulekha.com/news/feds-look-into-texting-before-deadly-train-crash.htm. Accessed May 29, 2009.

Associated Press. "Snapchat Lawsuit: Couple Claims App's 'Speed Filter' Led To Crash." *CBSBoston*, 2016, April 28. http://boston.cbslocal.com/2016/04/28/snapchat-lawsuit-speed-filter-crash/. Accessed July 23, 2017.

Associated Press. "Temporary Blindness Tied to Smartphone use in Dark." *Sarasota Herald-Tribune*, 2016, June 26. http://www.heraldtribune.com/news/20160626/temporary-blindness-tied-to-smartphone-use-in-dark. Accessed July 23, 2017.

Audi, T. "Rail Fans in Deadly Light: Subculture Includes Boy Who Texted Engineer in Crash." *Wall Street Journal*, 2008, September 19.

Axxidental. "How Is Your Phone Changing You?" *Axxidental Zine (Accidental Bear)*, 2016, June 2. http://accidentalbear.com/how-is-your-phone-changing-you-video-from-asapscience/. Accessed July 23, 2017.

Balsamo, Anne. *Technologies of the Gendered Body: Reading Cyborg Women*, Duke University Press, 1996.

Banet-Weiser, Sarah, and Laura Portwood-Stacer. "'I Just Want to be Me Again!': Beauty Pageants, Reality Television and Post-Feminism." *Feminist Theory*, vol. 7, no. 2, 2006, pp. 255–272.

Barad, Karen. *Meeting the Universe Halfway: Quantum Physics and the Entanglement of Matter and Meaning*, Duke University Press, 2007.

Bartky, Sandra Lee. "Foucault, Femininity, and the Modernization of Patriarchal Power." In *Feminism and Foucault: Reflections on Resistance*, edited by I. Diamond and L. Quinby, Northeastern University Press, 1988.

Bashford, Alison, and Claire Hooker. 2001. "Introduction: Contagion, Modernity and Postmodernity." In *Contagion: Historical and Cultural Studies*, edited by Alison Bashford and Claire Hooker, Routledge, 2001, pp. 1–12.

Bordo, Susan. *Unbearable Weight: Feminism, Western Culture, and the Body*. Tenth Anniversary Ed., University of California Press, 2003.

Borland, Sophie. "Don't Pokemon Go and Drive! More than 110,000 Road Accidents in the U.S. Were Caused by the Game in Just 10 Days." *Daily Mail*, 2016, September 16. http://www.dailymail.co.uk/sciencetech/article-3793050/Don-t-Pokemon-drive-110-000-road-accidents-caused-game-just-10-days.html. Accessed July 13, 2017.

Brody, Jane. "Screen Addiction Is Taking a Toll on Children." *New York Times*, 2015, July 6. https://well.blogs.nytimes.com/2015/07/06/screen-addiction-is-taking-a-toll-on-children/. Accessed July 13, 2017.

Browne, Kathe, and Catherine Nash, eds. *Queer Methods and Methodologies: Intersecting Queer Theories and Social Science Research*, Routledge, 2016.

Brustein, Joshua. "Has Oculus Finally Conquered the Virtual-Reality Vomit Problem?" *Bloomberg Technology*, 2014, December 4. https://www.bloomberg.com/news/articles/2014-12-04/has-oculus-finally-conquered-the-virtualreality-vomit-problem. Accessed Nov. 21, 2017.

Burke, Kenneth. *Permanence and Change: An Anatomy of Purpose*, University of California Press, 1984.

Cartwright, Lisa. *Screening the Body: Tracing Medicine's Visual Culture*, University of Minnesota Press, 1995.

CBA/AP. "'Exorcism' in Morgan Freeman Step-granddaughter's Death?" *CBS News*, 2015, August 17. http://www.cbsnews.com/news/exorcism-in-morgan-freeman-step-granddaughters-death/. Accessed July 13, 2017.

Chouliaraki, Lilie. *The Spectatorship of Suffering*, Sage, 2006.

Colossus: The Forbin Project. Dir. Joseph Sargent. 1970, USA.

Connell, R. "Metrolink Got a Waiver to Not Post Simple Safety Signs." *Los Angeles Times*, 2009, December 17. latimes.com/news/local/la-me-metrolink-signs17-2009dec17,0,5114265.story. Accessed December 18, 2009.

Conrad, Peter, and Joseph W. Schneider. *Deviance and Medicalization: From Badness to Sickness*, Temple University Press, 2010.

Contagion. Dir. Steven Soderbergh. 2011, USA.

Contrera, Jessica. "The Secret Lives of Screen-Addicted Teens." *The Week*, 2016, June 12. http://theweek.com/articles/629044/secret-lives-screenaddicted-teens. Accessed July 13, 2017.

Crist, Carolyn. "Leaning Forward During Phone Use May Cause 'text neck'." *Reuters*, 2017, April 14. http://www.reuters.com/article/us-health-neck-smartphone-use-idUSKBN17G12Z. Accessed July 13, 2017.

Cvetkovich, Ann. *Depression: A Public Feeling*, Duke University Press, 2012.

The Dark Knight. Dir. C. Nolan. 2008, USA.

De la Peña, Carolyn. *The Body Electric: How Strange Machines Built the Modern American*, NYU Press, 2005.

Dij, D. "Text Messaging May Have Caused Train Wreck that Kills 25 in California!" *TechRepublic*, 2008, September 13. http://techrepublic.com.com/5208-1035-0.html?forumID=102andthreadID=273818andstart=0. Accessed May 29, 2009.

Dr. Phibes Rises Again! Dir. R. Fuest. 1972, USA.

Dunn, J. D. "The Power of Christ Compels Him." *411mania*, 2008, July 1. 4http://www.411mania.com/politics/columns/79029/The-Power-of-Christ-Compels-Him.htm. Accessed May 12, 2012.

Estes, Adam C. "My Smartphone Gave Me a Painful Neurological Condition." *Gizmodo*, 2016, March 14. https://gizmodo.com/my-smartphone-gave-me-a-painful-neurological-condition-1711422212. Accessed July 13, 2017.

Feenberg, Andrew. *Questioning Technology*, Routledge, 1999.

Firger, Jessica. "Depression, Loneliness Linked to Binge-watching TV." *CBSNews.com* 2015, January 29. http://www.cbsnews.com/news/depression-loneliness-linked-to-binge-watching-television/. Accessed July 13, 2017.

Fischer, Claude S. *America Calling: A Social History of the Telephone to 1940*, Berkeley, University of California Press, 1992.

Foucault, Michel. "About the Concept of the 'Dangerous Individual' in 19th-century Legal Psychiatry." *International Journal of Law and Psychiatry*, vol. 1, 1978, pp. 1–18.

Foucault, Michel. "The Political Technology of Individuals." In *Technologies of the Self: A Seminar with Michel Foucault*, edited by Michel Foucault, Luther H. Martin, Huck Gutman and Patrick H. Hutton, University of Massachusetts Press, 1988, pp. 145–162. http://philpapers.org/rec/FOUTPT.

Foucault, Michel. "The Subject and Power." *Critical Inquiry*, vol. 8, no. 4, 1982, pp. 777–795.

Foucault, Michel. "19 March 1976." In *"Society Must Be Defended": Lectures at the Collége de France, 1975–76*, edited by Francois Ewald, M. Bertani, and A. Fontana, Picador, 2003, pp. 291–321.

Gallagher, James. "E-books 'Damage Sleep and Health,' Doctors Warn." *BBC News*, 2014, December 23. http://www.bbc.com/news/health-30574260. Accessed July 30, 2017.

"Gay Hispanic Train Engineer Crashes Train while Texting Young Boys, Kills 25." *NewsNet14.com*. 2008, September 17. http://www.newsnet14.com/2008/09/gay-hispanic-train-engineer-crashes-train-while-texting-young-boys-kills-25/. Accessed May 29, 2009.

Glassner, Barry. *The Culture of Fear: Why Americans Are Afraid of the Wrong Things*, Basic Books, 1999.

Gog. Dir. Herbert Strock. 1954, USA.

Gold, S. "Crash Shines Rare Light on 'Railfans': Train Hobbyists Fervent, Obscure." *Chicago Tribune*, 2008, October 5. http://www.chicagotribune.com. Accessed October 4, 2008.

Gold, S. "Train Hobbyists Are Loco for That Motion." *Los Angeles Times*, 2008, October 1. http://www.latimes.com/news/local/la-me-foamers1-2008oct01,0,5559030.story. Accessed October 1, 2008.

Governors Highway Safety Association, May 2013, http://www.ghsa.org/html/stateinfo/laws/cellphone_laws.html.

Halberstam, Judith. *Skin Shows: Gothic Horror and the Technology of Monsters*, Duke University Press, 1995.

Halloween. Dir. John Carpenter. 1978, USA.

Halloween. Dir. R. Zombie. 2007, USA.

Halsey III, A. "Hands-free Cell Phones Just as Likely to Distract Drivers." *Washington Post*, 2010. http://www.washingtonpost.com/wp-dyn/content/article/2010/01/29/AR2010012900053.html. Accessed February 25, 2010.

Hernandez, Vittorio. "Too Much Facebook Can Kill You." International Business Times, 2016, February 23. http://www.ibtimes.com.au/too-much-facebook-can-kill-you-1505749. Accessed November 21, 2017.

Hartouni, Val. *Cultural Conceptions: On Reproductive Technologies and the Remaking of Life*, University of Minnesota Press, 1997.

Hayles, N. Katherine. *My Mother Was a Computer: Digital Subjects and Literary Texts*, University of Chicago Press, 2005.

Heidegger, M. *The Question Concerning Technology and Other Essays*, Harper Torchbooks, 1977.

Hellraiser III: Hell on Earth. Dir. A. Hickox. 1992, USA.

Hennessy-Fisk, M., and S. Gold. "'Railfans' Create Virtual Memorial for Engineer." *Los Angeles Times*, 2008, September 19, B-1. http://articles.latimes.com/2008/sep/19/. Accessed November 25, 2008.

Horton, Helena. "More People have Died by Taking Selfies this Year Than by Shark Attacks." *The Telegraph*, 2015, September 22. http://www.telegraph.co.uk/technology/11881900/More-people-have-died-by-taking-selfies-this-year-than-by-shark-attacks.html. Accessed July 20, 2017.

"How to be Situationally Unaware." Episode of *How To Do Things*, 2014, April 2. Dir. Rachel Schrey. http://www.imdb.com/title/tt4537372/

Hutcheon, Linda, and Michael Hutcheon. *Opera: Desire, Disease, Death*, University of Nebraska Press, 1996.

Kardaras, Nicholos. "The Dangers of Pokémon Go: Kids' Brains Are Vulnerable to Virtual and Augmented Reality." *Salon*, 2016, August 11. http://www.salon.com/2016/08/11/the-dangers-of-pokemon-go-why-kids-brains-are-more-vulnerable-to-virtual-and-augmented-realitys-risks/. Accessed July 20, 2017.

KCBS Evening News. Dir. John Laxman. 2010, January 19, USA.

Kelley, Tristan. "Study: You are More Likely to Die Walking with Headphones." *Albany Times-Union*, 2012, January 17. http://www.timesunion.com/news/article/Study-You-are-more-likely-to-die-walking-with-2578662.php. Accessed July 20, 2017.

Kennedy, Michael. *A Brief History of Disease, Science and Medicine from the Ice Age to the Genome Project*, Asklepiad Press, 2004.

Kirby, Lynn. *Parallel Tracks: The Railroad and Silent Cinema*, Duke University Press, 1997.

Kulpa, Robert, and Mia Liinason. "Queer Studies: Methodological Approaches. Follow-Up." *Graduate Journal of Social Science*, vol. 6, no. 1, 2009, p. 1.

Lasch, Christopher. *The Culture of Narcissism*, W. W. Norton, 1979.

Lee, Adriana. "Warning: The iPhone 6 May Yank That Hipster Beard Right Off Your Face." *Read Write*, 2014, October 7. https://readwrite.com/2014/10/07/iphone-6-beardgate-hairgate-hipsters-beware/. Accessed July 20, 2017.

Leslie, Ian. "Google Makes Us All Dumber: The Neuroscience of Search Engines." *Salon*, October 12, 2014. http://www.salon.com/2014/10/12/google_makes_us_all_dumber_the_neuroscience_of_search_engines/. Accessed July 20, 2017.

Liinason, Mia, and Robert Kulpa. "Queer Studies: Methodological Approaches." *Graduate Journal of Social Science*, vol. 5, no. 2, 2008, pp. 1–4.

Linshi, Jack. "iPhone Separation Anxiety Makes You Dumber, Study Finds." *Time.com*, 2015, January 11. http://time.com/3662846/iphone-separation-study/. Accessed July 20, 2017.

The Living Daylights. Dir. J. Glen. 1987, USA.

Lööv, Anna Olovsdotter. "Queering Methodology." *NORA-Nordic Journal of Feminist and Gender Research*, vol. 20, no. 2, 2012, pp. 158–162.

Lupton, Deborah. *Medicine as Culture: Illness, Disease and the Body*, Sage, 2012.

Lutz, Tom. *American Nervousness 1903: An Anecdotal History*, Cornell University Press, 1991.

Marwick, Alice. To Catch a Predator? The MySpace Moral Panic. *First Monday*, vol. 13, no. 6, 2008. http://firstmonday.org/ojs/index.php/fm/article/view/2152/1966. Accessed June 6, 2013.

Mathis, T. "Homosexuality is a Dangerous Mental Disorder: Study: Homosexual Lifestyle Strongly Linked to Depression, Suicide." 2008, September 28. http://tedmathis.blogspot.com/2008/09/homosexuality-is-dangerous-mental.html. Accessed September 28, 2008.

Melloy, K. "Bereaved Gay Man at Controls in L. A. Train Crash." *Edge News*, 2008, September 17. http://www.edgeboston.com/index.php?ch=newsandsc=andsc3=andid=80494andpf=1. Accessed September 28, 2008.

Mikey. Dir. D. Dimster. 1992, USA.

"Million Dollar Abie," *The Simpsons* episode. Dir. Steven Moore. 2006, USA.

"Off the Rails." *Goodasyou.org*. September 28, 2008. http://www.goodasyou.org/good_as_you/2008/09/off-the-rails.html. Accessed May 19, 2011.

Oswald, Ed. "The 'Upgrade Effect' Is Why You Seem to Break Your iPhone When a New One's Out." *Digital Trends*, 2016, December 20. https://www.digitaltrends.com/mobile/iphone-upgrade-effect/. Accessed July 20, 2017.

Parsons, T. *The Social System*, Routledge, 1991.

Pascual, Katrina. "What Is 'Phubbing'? Here's How This Cellphone Habit Can Ruin Your Relationship." *Tech Times*, 2015, October 3. http://www.telegraph.co.uk/news/health/news/12167957/Mobile-phones-are-cooking-mens-sperm.html. Accessed July 20, 2017.

Pelling, Margaret. 2001. "The Meaning of Contagion: Reproduction, Medicine and Metaphor." In *Contagion: Epidemics, History and Culture from Smallpox to Anthrax*, edited by Alison Bashford and Claire Hooker, Pluto Press, 2002.

Pernick, Martin S. 2002. "Contagion Culture." *American Literary History*, vol. 14, no. 4, pp. 858–865.

Quinones, S., and D. Kelley. "Metrolink 111 Engineer Led Solitary Life Marred by Tragedy." *Los Angeles Times*, 2008, September 18. http://www.latimes.com/news/printedition/front/la-me-engineer17-2008sep17,0,6452601.story. Accessed September 28, 2008.

Rafael, Vincente L. "The Cell Phone and the Crowd: Messianic Politics in the Contemporary Philippines." In *New Media Old Media: A History and Theory Reader*, edited by Wendy Hui Kyong Chun and Thomas Keenan, Routledge, 2006.

Rettner, Rachael. "Man Tears Tendon After Playing Candy Crush for Weeks." *LiveScience*, 2015, April 13. https://www.livescience.com/50466-smartphone-game-injury.html. Accessed July 15, 2017.

Rheingold, Howard. *Smart Mobs: The Next Social Revolution, Transforming Cultures and Communities in the Age of Instant Access*, Perseus, 2002.

Risling, G. "Metrolink Engineer's Life Held Challenges." *Ventura County Star*, 2008, September 18. http://www.venturacountystar.com/news/2008/sep/18/metrolink-engineers-life-held-challenges/. Accessed September 28, 2008.

Rosengarten, Marsha. "A Pig's Tale: Porcine Viruses and Species Boundaries." In *Contagion: Historical and Cultural studies*, edited by Alison Bashford and Claire Hooker, Routledge, 2001, pp. 168–182.

Sahagun, L., and T. Rohrlich. "Texting Eyed in Fatal Train Collision." *Chicago Tribune* 2008, September 15. http://archives.chicagotribune.com/2008/sep/15/nation/chi-la-trainsep15. Accessed May 29, 2009.

Salinger, Tobias. "Selfies Lead to 'Social Media Lice'." New York Daily News, 2015, August 25. http://www.nydailynews.com/life-style/health/selfies-lead-social-media-lice-experts-article-1.2337345. Accessed November 21, 2017.

Saturn 3. Dir. Stanley Dorren/John Barry. 1980, UK.

Scary Movie 4. Dir. Daryl Zucker. 2006, USA.

Sellers, Patrice. "Are You Situationally Aware?" *Fortune*, 2009, October 28. http://fortune.com/2009/10/28/are-you-situationally-aware/. Accessed July 23, 2017.

Serlin, David. *Replaceable You: Engineering the Body in Postwar America*, University of Chicago Press, 2004.

Shah, Sonali. "2009: The Year of SMS, Again." *Wireless Week*, January 29, 2009. http://www.wirelessweek.com/articles/2009/01/2009-year-sms-again. Accessed June 6, 2013.

Skultans, V. "An Historical Disorder: Neurasthenia and the Testimony of Lives in Latvia." In *Colonialism and Psychiatry*, edited by D. Bhugra and R. Littlewood, Oxford University Press, 2001.

Streitfeld, David. 2013. "Google Glass Picks Up Early Signal: Keep Out." *The New York Times*, May 6, sec. Technology/Personal Tech.

Subramania, Ashwin. "Tinder, Grindr, Other Casual Sex Apps Blamed for Skyrocketing STD Rates." *NewsEveryday*. 2015, May 30. http://www.newseveryday.com/

articles/18214/20150530/casual-sex-apps-blamed-rise-stds-report.htm. Accessed July 30, 2017.

Taussig, Michael T. 1980. "Reification and the Consciousness of the Patient." *Social Science & Medicine. Part B: Medical Anthropology*, vol. 14, no. 1, pp. 3–13.

"Texting + Driving = Lunacy." *Staten Island Advance*, 2008, September 24. http://www.silive.com/editorials/advance/index.ssf?/base/news/122225671028150.xmlandcoll=1. Accessed May 29, 2009.

Timberg, Scott. "How Video Games Suck You In: 'Our Sense of Time Becomes Yoked, Not to the Ticking of the Clock, But to the Pattern of our Interactions'" *Salon*, 2016, August 13. http://www.salon.com/2016/08/13/living-and-dying-the-best-games-present-a-reality-that-engages-us-totally-flatters-us-by-requiring-our-presence-and-input/. Accessed July 30, 2017.

Totten, Kristy. "Kill Shot: Selfies Killed More People Last Year Than Mount Everest." *Las Vegas Weekly*, 2016, February 10. https://lasvegasweekly.com/news/2016/feb/10/kill-shot-selfies-killed-more-people-last-year-tha/. Accessed July 30, 2017.

Turner, Bryan S. *The Body and Society: Explorations in Social Theory*, Third ed., Sage, 2008.

Untraceable. Dir. G. Hoblit. 2008, USA.

Van Dijck, José. *The Transparent Body: A Cultural Analysis of Medical Imaging*, University of Washington Press, 2005.

Varones, W. C. "KFI: Metrolink Crash Engineer was Psycho Gay Killer." 2008, September 28. http://wcvarones.blogspot.com/2008/09/kfi-metrolink-crash-engineer-was-psycho.html. Accessed May 19, 2011.

Virus. Dir. John Bruno. 1999, USA.

Weber, Brenda R. "Makeover as Takeover: Scenes of Affective Domination on Makeover TV." *Configurations*, No. 15, 2007, pp. 77–99.

Weber, Brenda R. "What Makes the Man? Television Makeovers, Made-over Masculinity, and Male Body Image." *International Journal of Men's Health*, vol. 5, no. 3, 2006, pp. 287–306.

Wood, D. B. "LA Metrolink Crash puts Focus on Dangers of Texting." *Christian Science Monitor*, 2008, September 17.

Zargar, Arshad. "Major City Sets No-selfie Zones After Spike in Deaths." *CBS News*, 2016, February 25. http://www.cbsnews.com/news/mumbai-sets-no-selfie-zones-after-spike-in-deaths-linked-to-selfies/. Accessed July 14, 2017.

· 1 ·

PATHOLOGY SHAPES SUBJECTS

Gendering and Normalizing

A tabloid in the 1980s reported that "4 Housewives in 5 Suffer Ailment Caused by Women's Lib."[1]

> Housewife syndrome—a tragic ailment that strikes women who have chosen not to join the working world—is a major health problem today, experts say. Its symptoms include depression and feelings of inferiority, as well as physical aches and pains—and experts say it is the fault of "fanatic" feminists who make housewives feel worthless and unappreciated.

In examples such as this, sickness defines and labels people along multiple dimensions. Housewives are mentally and physically unwell, reinforcing stereotypes of women as feeble, weak, and housebound. Feminists, in turn, are also categorized as hyperbolic, irrational "fanatics"—reinforcing the feminine stereotype of the hysterical harpy. However, they are also the carriers of disease: they infect innocent housewives with their malaise. Moreover, both groups bear responsibility. Feminists "make housewives feel worthless and unappreciated," yet housewives "have chosen not to join the working world," making themselves vulnerable to this infection. Nowhere in the article is any mention of societal or external causes. Housewives do not feel badly because a variety of social forces hinder their success in the working world, devalue the work they do at home, or rob them of independence.

In this chapter, I argue that pathologization categorizes along dimensions of health, but also of normality in general and gender in specific. Through an examination of historic medical texts and popular representations of the archaic diagnosis and proto-technopathology of neurasthenia, I will demonstrate the role of pathologization in shaping subjects. I will then show the continued salience of these discourses, even as the diagnosis disappears and neurasthenia evolves into schizophrenia, focusing on the transitional moment of the 1947 Joan Crawford film *Possessed*.

The Work of Disease: Neurasthenia's Normalizing and Gendering

In maintaining an agential realist perspective, I focus on disease as action, not thing. Pathologization establishes oppositional relationships. Sander Gilman describes it as "a central marker for difference"—creating opposition between Self and Other.[2] Furthermore, the individual body frequently operates as a symbol of the social body.[3]

Pathologization not only creates differences, it assigns them values. As Bryan Turner describes, "The concepts of 'illness,' 'disease,' and 'health' inevitably involve some judgment which ultimately rests on a criterion of statistical frequency or an ideal state. The 'average individual' does not exist. … Disease is not a fact, but a relationship and the relationship is the product of classificatory *processes*."[4] Disease can even categorize as good when it is a marker of positive social distinction.[5] From a disability studies perspective, demarcating the abnormal defines and maintains normalcy, and intersects with other systems of differently valuing bodies, such as race, nationality, sexuality, and gender.[6] More typically, however, pathologization, and related medical discourses and technologies, reinforce existing social hierarchies through justifications of medical inferiority or deviance.[7] Gilman writes, "The very concept of pathology is a line drawn between the 'good' and the 'bad',"[8] and familiar examples include racist medical technologies from eugenics to evolutionary physiognomy, imperialist colonization of "dirty savages," or, infamously, enslaved Africans in the United States diagnosed with the mental illness of drapetomania—an "irrational" desire to run away from their masters. A century later, during the civil rights movement, schizophrenia was associated with race such that one state hospital for the criminally insane experienced a sudden, atypical increase in black male patients, all diagnosed with schizophrenia.[9] Concurrently, a tabloid reported on "ghetto headache," quoting a doctor that, "As little as 10 years ago, ... the ghetto resident was better adjusted to his fate, but the recent awakening of social consciousness has made him yearn for a better life."[10]

Pathologization is, then, part of a broader process of hierarchical normalization. Also, it is part of, or, to use agential realist terminology, entangled with the process of gendering—associating persons, their attributes, and actions with degrees of masculinity and femininity. There is a history of associating femininity with disease, and, therefore, pathologization is specifically a process of feminization. Feminist scholars have shown how normal functions of the female body or psyche have been pathologized, reinforcing male privilege, such as Maines' history of the vibrator as, not an aid to female orgasmic pleasure, but a medical device for physicians to stimulate sick women into therapeutic convulsions.[11] Laqueur argues for pathological roots at the very understanding of sex and gender, demonstrating the history and ongoing salience of a "one-sex model" in which women (and the feminine) are considered a lesser, abnormal version of men and masculinity, rather than an opposite.[12] Gilman writes, "Pathology is disorder and the loss of control, the giving over of the self to the forces that lie beyond the self."[13] In sickness, a person is *de*masculinized to some degree through their loss of bodily control, physical integrity, mental acuity, and/or individual agency.

This is not an absolute process of fixed and tidy mapping (i.e., "sick" = "feminine"). It is dynamic and protean, and there are exceptions: feminine pathologies can be idealized (Romantic-era consumptives, contemporary anorexics), and pathologies can be masculine (sports injuries, hypertension). Every marking and categorization of less masculine or more feminine does not permanently or unequivocally signify disease or abnormality. But neither are these process unrelated.

Normalization and gendering are evident in the archaic disease of neurasthenia, a proto-technopathology diagnosed during a transitional moment in the separation of biological and mechanical energies, which I will address in a subsequent chapter. In 1869, neurologist George Beard conceptualized the disease as literally nerve (*neuro*) + weakness (*asthenia*), an exhaustion of the bioelectrical nervous system by overstimulation.[14] This diagnosis, no longer used in the West, has its most familiar vestigial form in the concept of a nervous breakdown. Neurasthenia, however, was historically a legitimate and popular diagnosis. Beard conceived of it as a particularly American phenomenon, due to perceived climatological uniqueness and the competitiveness and ambition that came with the ideology of American exceptionalism.[15] It would, however, spread to global usage for decades beyond its use in the United States.

Well-known neurasthenics included Max Weber, Theodore Roosevelt, and Theodore Dreiser. It was referred to as "the blues" before that term had a musical meaning, but the physical and mental manifestations of an overstimulated nervous system went beyond the contemporary meaning of "the blues" as depression. Neurasthenia was nearly unbounded in its symptomology, a catch-all for mental and physical disorders, such as dyspepsia, extreme nervousness, lethargy, hyperexcitability, insomnia, impotence, skin rashes, depression, hay fever, liver trouble, headaches, premature baldness, inebriety, spermatorreah, epilepsy, and mild insanity. Neurasthenia was also progressive, leading to other conditions, as Beard charted in his "Evolution of Nervousness":

General Nerve Sensitiveness

↓

Nervous Diathesis (nervous dyspepsia → sick-headache → near-sightedness → Chorea → sleeplessness → Asthenopia → hay-fever → hypochondria → hysteria)

↓

Nervous Exhaustion (Neurasthenia) (cerebral → spinal → digestive → sexual)

↓	↓	↓
inebriety	insanity	epilepsy[16]

"The chief and primary cause of this development and very rapid increase of nervousness," wrote Beard, "is *modern civilization*, which is distinguished from the ancient by these five characteristics: steam-power, the periodical press, the telegraph, the sciences, and the mental activity of women."[17] Technologies of transportation and communication, coupled with a challenge to gender roles, were "at the head" of "predisposing causes" of neurasthenia. Neurasthenia was linked with modern civilization, but, importantly, it was a physical disease of organic nature, *not* a psychosomatic or mental illness as we might think of today, such as stress. As Beard described:

> None of the predisposing or exciting causes that follow it are competent to produce functional nervous disease of the class described in this work unless civilization prepares the way. The nervous diseases from which savages suffer, and the lower orders of peasantry, are largely of a subjective, psychical character, being caused by the emotions, and assume very different phases, whatever their names be, from those herein described.[18]

Here we see pathologization's normalizing work, marking ideals of race and class. Neurasthenics were modern, American (or Americanized), and rational. Neurasthenia was *not* mental illness, i.e., "of a subjective, psychical character, being caused by the emotions." Neurasthenia exempted its sufferers from the moral and character failings associated with mental illness at the time, demonstrating how pathologization can have positive social values as well. Indeed, neurasthenia afflicted upper-class "brain workers," not manual laborers. However, Beard had to perform some categorical twists to maintain ideals of civilization as both noble and corrupting: "savages" suffered nervous diseases caused by emotions, but not physical or "functional" ones of the type that impacted the white upper class. Citing studies of emancipated Negroes on islands off the coasts of Charleston and Savannah—"a bit of barbarism on our door-steps"—Beard argued that, "there is almost no insanity among these Negroes; there is no functional nervous disease or symptoms among them."[19] Instead, stereotypical attributes of laziness and sloth were reinforced and medically explained as due to their *lack* of modern, Caucasian, nervous energy.

Other authorities blended mental and physical understandings of neurasthenia. American psychologist Morton Prince wrote in the 1890s that, although a "real" organic disease may be a trigger, the patient's obsessive, irrational fixation upon it expands the condition to neurasthenic, hysterical problems. What we would think of today as hypochondria or psychosomatic illness, or specific examples, such as "hysterical pregnancy," he described as "neuro-mimesis," the brain imitating the symptoms of a real affliction. "That pain and various other nervous phenomena may be kept alive by the mind dwelling on it, is a well-known fact."[20] Moreover, in a separate but contemporaneous essay, he wrote that the treatment is largely educational: the physician, acting as a guide to the patient in terms of morality and character, also teaches them to separate real from obsessive suffering.[21]

As these descriptions should suggest, neurasthenic discourses contributed to and supported stereotypes and other discourses of gendering. By some accounts, more women were diagnosed with neurasthenia than men. Neurasthenia was also seen as indicative of the inherent degeneracy and susceptibility of women to disease.[22] For example, an 1899 issue of *Medical News* cited "the sensitive dispositions of delicate girls during the development period" as making them vulnerable to neurasthenia, chlorosis, and neurotic conditions.[23] Neurasthenia was opportunistic, taking advantage of the laxities in morality and character that women, people of color, and lower classes were seen as having. While modern society's demands could sicken the elite, non-elites were susceptible to neurasthenia due to their already-weak, naturally susceptible constitutions.

Neurasthenic femininity was not "woman" in an essential state, but a specifically modern femininity, intersectionally raced and classed. Beard wrote, "Woman in the savage state is not delicate, sensitive, or weak. Like man, she is strong, well developed, and muscular, with capacity for enduring, as well as childbearing. The weakness of woman is all modern, and it is preeminently American."[24] The modern society that caused neurasthenia was a feminizing over-civilization, an occupational hazard of progress. It was abnormalizing and yet, as suggested previously, a marker of distinction. Neurasthenia feminized, but in a modern, white, upper-class way that was valued above "savage" femininity.

The disease feminized men. In 1871, William Dean Howells disparaged the epidemic of neurasthenia among artists and intellectuals by writing that "'sometimes America seems little better than a hospital for invalid women'."[25] S. Weir Mitchell, inventor of the neurasthenic "rest cure," wrote in 1877, "I have many a time seen soldiers who had ridden boldly with Sheridan or fought gallantly with Grant become, under the influence of painful nerve-wounds, as irritable and hysterically emotional as the veriest girl."[26] Such associations continued into Europe, where, by the 1930s, a Germanic concern with threats to masculinity, modern exhaustion, and depleted willpower was associated with a backlash against feminism.[27]

Pathological gendering leads to classifying and valuing sexualities. One cultural history of male impotence argues that neurasthenia was a focal point in gendered cultural anxieties: its nervous exhaustion was due to the emasculating and feminizing stresses of modern civilization, which also led to medicalized failures of masculinity such as perversion, overindulgence, masturbation, and homosexuality.[28] Freud's analysis of Daniel Paul Schreber, the author of *Memoirs of my Nervous Illness*, focused on his homosexuality, and neurasthenia historian Tom Lutz argued that "homosexuality also would have created the kind of role and status anxiety so often expressed in neurasthenia."[29] American gay rights pioneer Harry Hay recalled how, in the 1920s, there was no word "gay" to describe homosexual men: "The word for us was 'temperamental'."[30] "Temperamental" parallels the emotional liability and moodiness of neurasthenia, and its pathological (ef)feminization. Insatiable heterosexual demands of hysterical women also contributed to heterosexual male neurasthenic exhaustion. A traumatic sexual humiliation, such as inability to satisfy the demands of an insatiable wife or voracious mistress, could push nerves already over-excited by modern life over the edge into neurasthenia.

Biological sex and gender ideals were central to treatments for neurasthenia. Male neurasthenics needed invigorating, outdoor activities to amp up their masculinity; female neurasthenics, exhausted by forays into modern public spheres, needed a recuperative escape for their ravaged, weaker systems. And yet, despite having reorganized some character failings into the more acceptable realm of physical illness, stereotypes of inherent feminine weakness persisted. Rejecting a school of thought that posited a nutritional cause of neurasthenia, American psychologist Morton Prince told the Massachusetts Medical Society in 1898 "that the cures effected by the rest treatment are due to other influences, largely moral and educational."[31] Note that he specified only the neurasthenics taking the rest cure, which were typically women. Neurasthenic men who took to the hills and horseback riding were apparently more legitimate, innocent victims of a real disease.

Medical records from 1870 to 1910 show that, despite what might have been original functions of neurasthenia in pathologization along gender, class, and race, the condition democratized fairly rapidly.[32] Rural, urban, and small-town physicians (not just elite neurologists) came to diagnose and treat Americans of all social classes and types of employment in homes, offices, public clinics, private sanitariums, city hospitals, and state asylums across the country. For these people as well as elites, a respectable, even admirable diagnosis of neurasthenia served to replace, cover, or augment previous socially undesirable diagnoses, such as hysteria, hypochondria, melancholia, and even syphilis. W. E. B. Du Bois was not neurasthenic but familiar with it from his teacher William James, and appropriated its symptoms to describe his famous concept of African-American double-consciousness, rhetorically realigning the racial category with a more socially esteemed class category.[33]

Lutz argues that such flexibility was key to neurasthenia's popularity and eventual downfall. One prominent psychologist noted as early as 1890 that "neurasthenia, as a term, means very little."[34] Diagnoses of neurasthenia went on the decline, displaced to some degree by the emergence of Freud and also perhaps because, after the shock of World War I, the disease's association with cultivated refinement and sensitive temperaments was less tenable. By 1923, one physician wrote, "In neurasthenia no symptoms have hard-and-fast lines."[35] A prominent Scottish physician felt that, after World War I, rest cures no longer worked for nerves, and realigned the diagnosis with character failings and mental illness: "Foolishness untreated (and not, as scientists used to believe, chemical, physical, or bacteriological ongoings in the brain) is the main cause, not merely of our neurasthenia and nervous breakdowns, but of our rising insanity rate."[36]

In 1923, a prominent UK physician looking back at nervous afflictions before, during, and just after World War I, argued that neurasthenia was a moral failing, much like alcoholism and venereal disease. Rejecting as faddish the then-new germ theories of bacteriology, he asserted that nervous diseases seldom had an organic basis:

> We need hardly take the biological factors into account; the causes were rather of a social and moral order; it was here a case of interaction between psyche and society rather than between psyche and bodily processes. … It was not so much the circumstances as the reaction to circumstances that was at fault here.

Moreover, although poor education was a factor, these reactions are at times framed not merely as the fault of the victim's inherent condition, but the victim's active *choice*. "Ennui or boredom passed easily into Nervous Debility—from discomfort into actual pain. Now, this ever-increasing pain of the pre-war period … arose very largely from the fact that people *were allowing themselves* to become weak."[37]

Representations of Neurasthenia: O. Henry and Charlotte Perkins Gilman

The phenomena of neurasthenia traversed popular as well as medical culture. As Karen Barad reminds us, representations are not separate things from what they represent, but mutually constitutive, what she calls material-discursive practices. I now will describe examples of the normalizing and gendering work of neurasthenia seen in two literary examples.

In 1910, American humorist O. Henry published a brief illustrated story, *Let Me Feel Your Pulse: Adventures in Neurasthenia.* It relates a first-person account of a man's treatment at a sanatorium. Neurasthenia's manifold treatments are parodied in a running joke about the oxymoronic prescription of "absolute rest and exercise."[38] Equally ripe for humor are the generally less-than-masculine fellow patients (a female novelist, a Wall Street broker, an architect, a minister, society ladies, a neurotic millionaire, and a famous artist), the inauthentic disease itself (suggested to be merely a cover for alcoholism), and the indulgently romantic sufferer himself: "There is nothing more alarming to a neurasthenic than to feel himself growing well and cheerful."[39]

The story's conclusion makes explicit that the disease is one of failed heterosexual masculinity. His doctor tells the protagonist that a magic cure-all mountain plant is the only thing that can save him. They search for it for a month. Finally, when the doctor spies the protagonist with a female friend, Amaryllis, he takes her aside. The doctor tells her to look up her name in an encyclopedia when they get home and tell the protagonist what it means.

"It seems to be the name of a genus of flowering plants and also the name of a country girl in Theocritus and Vergil," she tells him. "What do you suppose the doctor meant by that?" (Fig. 1.1)

> "I know what he meant," said I. "I know now."
>
> A word to a brother who may have come under the spell of the unquiet Lady Neurasthenia.
>
> The formula was true. Even though gropingly at times, the physicians of the walled cities had put their fingers upon the specific medicament.
>
> And so for the exercise one is referred to good Dr. Tatum on Black Oak Mountain—take the road to your right at the Methodist meeting-house in the pine-grove.
>
> Absolute rest and exercise!
>
> What rest more remedial than to sit with Amaryllis in the shade, and, with a sixth sense, read the wordless Theocritan idyll of the gold bannered blue mountains marching orderly into the dormitories of the night?
>
> The End.[40]

"Amaryllis" is not only, as suggested, a name for a shepherdess, country girl, or rural sweetheart in classical and pastoral poetry, but also a flowering plant (the "magic cure-all") known for its red and pink trumpet or funnel-shaped blossoms, also known as a "naked lady." The Romantic fad for a symbolic language of flowers, enabling secret messages and difficult sentiments to be expressed through arrangements, had recently been revived at the time of O. Henry's writing, with red amaryllis meaning, "I respect you from the depths of my soul." Thus, O. Henry clearly seems to be asserting repaired masculinity, through heterosexual love and sexuality, as the "cure-all" for the dubious condition of male neurasthenia.

A more famous example, Charlotte Perkins Gilman's short story, "The Yellow Wall-paper," is a literary classic and one of the most famous feminist critiques of gendered medical practice.[41] In her story, the enforced rest cure for

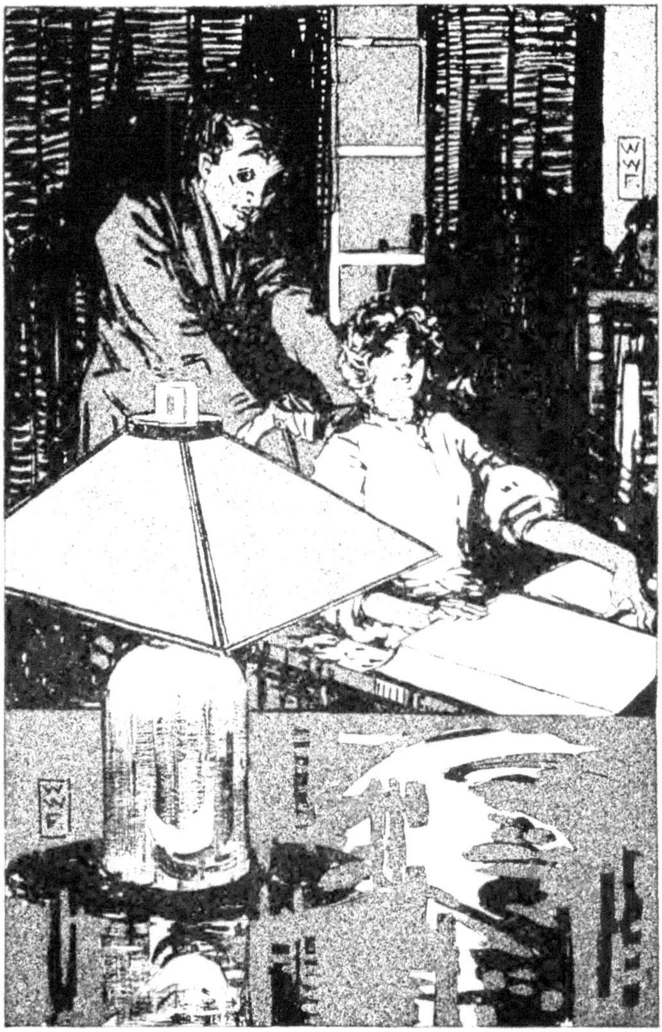

"What do you suppose the doctor meant by that?"

Figure 1.1. "What do you suppose the doctor meant by that?" The prescription for male neurasthenia is heterosexuality.
Source: From Henry, O. *Let Me Feel Your Pulse: Adventures in Neurasthenia*. New York: Doubleday, Page and Co. 1910. Book by O. Henry, illustration by W. W. Faucett. Internet Archive. Public domain. https://archive.org/details/letmefeel00henrrich.

female neurasthenics reinforces domestic servitude, exhaustion, and exclusion from intellectual society and activity, because they supposedly caused nervous illness in the first place. Originally published 1892, it was considered a horror tale in the vein of Edgar Allan Poe, and never truly lost, but anthologized several times. However, its 1973 republication by Feminist Press, with an afterword by Elaine R. Hedges, ushered in unprecedented international acclaim and led to it becoming one of the most-studied English texts.

Briefly, the story concerns a mostly unnamed female protagonist who suffers from nervous illness, but her physician-husband does not believe she is sick, in an early example of Joseph Dumit's "illness you have to fight to get."[42] She takes the Mitchell rest cure and is isolated in a room with decorative wall-paper. The forced inactivity makes her symptoms worse, and she becomes drawn to the wall-paper's design, studying it, feeling nauseated by its colors, detecting a smell, and seeing a pattern in it. She grows paranoid toward other characters and realizes the pattern within the design is a woman, who begins to seem mobile or sentient. In an ambiguous ending, her husband John arrives and asks what is the matter. She responds: "I've got out at last … in spite of you and Jane. And I've pulled off most of the paper, so you can't put me back!"[43] John faints, and she crawls over his supine body repeatedly. It is unclear what the ending means: Has the wall-paper woman has taken over the protagonist, or was she successfully refused? Was Jane (never before named) the protagonist? Is she now liberated from "Jane's" domestic life? Liberated, trapped, triumphant, or insane?

As clarified in its prequel, "Through This," she was not sickened by education, technology, or access to her husband's books (as one contemporary doctor blamed postpartum distress). Contrary to medical opinion at the time, her volatile uterus and reproductive organs are never mentioned as cause, either. As the story suggests, and its prequel makes clear, it is exhaustion from domestic responsibilities. "Through This" is Gilman's brief portrait of a harried wife and mother, featuring most of the characters in "Wall-paper." The protagonist is too exhausted at the end of the day to respond to a letter from Jane, who is facing challenges as a new mother. The story, using a different case, is a "Before" portrait to the "After" of "Wall-paper."

Furthermore, the institutionalized sexism of the medical establishment is emphasized in Gilman's autobiography, in which she describes the target of "Wall-paper," Mitchell, scornfully telling her, "I've had two women of your blood here already"—referring to having treated Harriet Beecher Stowe's daughter Giorgiana Stowe (Gilman's great-aunt) and another unidentified Stowe woman.[44] "Blood" serves not only as vernacular for relations but also medicalizes feminist activism, conflating disease with politics and pathologizing not merely women, but a specific type of women: feminists.

"Wall-paper" also links neurasthenia and gendering with the supernatural, suggesting the intertwined cultural histories of electricity, femininity, vital energies, and mediumship, to be discussed further in a later chapter. Briefly, there was thought to be an electro-psychic connection among nineteenth-century spiritualism, which expressed a gendered pathology: mediums were typically female and the most famous, the Fox sisters, were forced into asylums for treatment in the emerging field of psychiatry—a medical literal taming of the (often pubescent, sexually awakening) feminine electro-spiritual.[45] Gilman begins her story discussing the estate housing the room as "haunted" and her condition as her "ghostliness." Her husband, John, explicitly does not believe in the supernatural:

> John is practical in the extreme. He has no patience with faith, an intense horror of superstition, and he scoffs openly at any talk of things not to be felt and seen and put down in figures.
>
> John is a physician, and *perhaps*—(I would not say it to a living soul, of course, but this is dead paper and a great relief to my mind)—*perhaps* that is one reason I do not get well faster.
>
> You see he does not believe I am sick!

Note the quick progression from disbelief in the supernatural, to her lack of recovery, to disbelief in her illness.[46] Later in the story, the lady in the paper could be a ghost or hallucination, but the story significantly refuses to differentiate or specify the two.

Moreover, her story also foreshadows key elements of technopathologies: it brings together a sickening form of mechanical communication media (paper, writing, decoration) with a disease of bioelectricity and modern technological society (neurasthenia). The act of communication and its media are central and, although they are not electric machines, electricity still suffuses the story in its focus on depleted nervous energies. The story itself is an act of communicating back to Mitchell—an early example of interactive, participatory media perhaps—and to other women and doctors. This act is amplified by the prequel, essay, and autobiography, making clear that, if the story was meant to convey a subjective psychic experience, Gilman also wished to make the intended message of the story clear. Communication is also shown to be inappropriately faulted, when, in actuality, it is the cure: "John has cautioned me not to give way to fancy in the least. He says that with my imaginative power and habit of story-making, a nervous weakness like mine is sure to lead to all manner of excited fancies, and that I ought to use my will and good sense to check the tendency. So I try."[47] John's sister also is accused of thinking that writing makes the protagonist sick. And yet:

> I don't know why I should write this.
> I don't want to.
> I don't feel able.
> And I know John would think it absurd. But I must say what I feel and think in some way—it is such a relief![48]

This reversal of conventional wisdom parallels that of entire story: the "cure" makes you sick; the "cause" can cure you.

Most fascinating, however, is the communicative medium of the wall-paper. Yes, there has been much scholarly argument for preserving that hyphen, as it emphasizes that it is not merely a wall covering, but paper on the wall, the prime communication medium for centuries mounted on the boundaries of domestic imprisonment, her bedroom's "four walls." Typically one differentiates, however, between decorative patterns and linguistic symbols, yet Gilman challenges this distinction. As Treichler has discussed, the protagonist's confrontation is with language and discourse itself as techniques of patriarchal control and power. Language forms both sentences: the protagonist's prescribed sentence to her room for a rest cure and sentencing women to social subordination.[49] The supposedly mere aesthetics of the wall-paper do contain a message. "I never saw so much expression in an inanimate thing before." They do contain encoded or mediated human consciousness and agency. "There are things in that paper that nobody knows but me, or ever will." There is a woman in the pattern. The pattern is not mere decoration but is communication, challenging the gender roles associated with decoration and pragmatism, respectively. The wall-paper is animate and inanimate. Its liminality is sickening. The "sprawling flamboyant patterns committing every artistic sin" are described variously as confusing the eye, irritating, "outrageous," suicidal (not provoking suicide but suicidal themselves, again suggesting agency in decoration), "repellent, almost revolting; a smouldering unclean yellow," "lurid," "a sickly sulphur tint," "horrid," impertinent, "absurd," "irritating," "strange, provoking," "silly and conspicuous," "pointless," "bloated," "debased," "interminable," "hideous," "unreliable," "infuriating," "torturing," reminiscent of a fungus, strange, "foul," "tiresome and perplexing," dizzying, "awful," and smelling "yellow." (Lutz also notes the use of toxic dyes in Victorian wallpaper, adding another layer to this.) It provokes sickness, is sick itself, and even suggests a representation of ill health: "There is a recurrent spot where the pattern lolls like a broken neck and two bulbous eyes stare at you upside down." It changes with changing light, suggesting semiotic instability.

Indeed, as we will see, while "neurasthenia" as a signifier would give way to new terms, such as "schizophrenia," the processes of pathologization, gendering, and normalization would persist.

From Neurasthenia to Schizophrenia: A Woman *Possessed*

A transitional moment in neurasthenia's evolution is articulated in the 1947 Oscar-winning Joan Crawford film, *Possessed*. The film opens with a catatonic woman, Louise, found wandering the streets of Los Angeles. Hospitalized and medicated, she explains to doctors, as the film unfolds in flashback, her story of obsessive love for and rejection by a former lover. Louise was a home-care nurse, dumped by her caddish paramour, David. After Louise's female patient, a paranoid invalid, died, Louise had remained with the family, eventually marrying widower Dean. However, she was unable to let go of her passion for David who, coincidentally, worked for Dean. David instead began romancing Louise's new stepdaughter, driving Louise into madness. Hallucinating and paranoid, Louise became convinced she murdered Dean's first wife, who was her patient. In a jealous rage, Louise then murdered David. Her prognosis at the film's end is bleak, and contingent upon Dean's steadfast love and support.

In *Possessed* we see the normalizing and feminizing work of disease, but the disease in question unclear. Only one character is explicitly labeled as neurasthenic: Louise's patient. Early in the film, we see Louise working as a nurse for Pauline Graham, Dean's invalid wife who, although only heard, never seen, is clearly a "nut job." Neurasthenic Pauline foreshadows Louise's fate in several ways: she is bedridden, married to Dean Graham, paranoid and delusional with a "persecution complex," and fixating her delusions on a romantic love object. (Pauline mistakenly thinks Louise is having an affair with Dean; Louise is obsessed with David and, in the original script, has an affair with him while married to Dean.) Moreover, Pauline commits suicide, as does Louise in the original short story upon which the film was based; in the film, she merely takes on responsibility for Pauline's death.

Louise's diagnosis is contradictory. A psychiatrist she sees privately states that she exhibits "all the symptoms of neurasthenia." Yet, when Louise guesses that she has schizophrenia, he quickly confirms that diagnosis, which many of the film's press materials also mentioned. Although schizophrenia, as a mental condition, was conceptualized differently from the physical illness of neurasthenia, it was also a diagnosis that eventually absorbed many cases that physicians in previous decades would have diagnosed as neurasthenia. Therefore, although Louise was not a case study in neurasthenia, the film clearly links it to her parallel schizophrenia, both featuring catatonia, hallucinations, confusion, fatigue, emotional volatility, and obsession. Neurasthenia was already a somewhat archaic diagnosis in the West by 1947.[50] The film, then, is not a portrait of neurasthenia, but of the archaic diagnosis persisting in popular imagination and then-contemporary medical science.

Possessed presents a clear example of the pathologized feminine, especially feminine sexuality and romantic desire. The film is about the threat of the unrestrained, sick, obsessive love experienced by women, cited repeatedly and without qualification in the film's promotional materials and press coverage. This is further emphasized by press stories that blurred the line between Crawford and her character: Louise is "lovelorn;" Crawford is "thrice-divorced" and "swearing off matrimony." Many articles speculated on Crawford's difficult love life and concurrent divorce. Louise is ill; Crawford has a case of the sniffles during filming. Immediately after filming the climactic scene in which Louise shoots the object of her unrequited love, Crawford has to go to the doctor for a shot of penicillin. Director Curtis Bernhardt tells the press how Crawford had lost herself in the emotional intensity of shooting a hysterical scene: "The cameras stopped, the scene was finished, and I yelled: 'Cut. Stop, Miss Crawford. It's all over.' But she could not stop. She kept on screaming. Finally, we managed to get her back to her dressing room and she calmed down." The doctor in the film cites Louise's similar inability to distinguish reality and fantasy, a blurring emphasized by the film's use of innovative subjective cinema techniques, making the audience unclear what was and was not real for Louise, just as the media discourse around the film blurred what was and was not real for Crawford.[51]

Institutional structures of the film industry at the time also served to heighten the film's feminine pathologization. In early versions of the script, it was made fairly clear that David resumed his affair with Louise after she married Dean. The Motion Picture Association of America, in several memos to studio head Jack Warner, insisted that these references to adultery be removed. The result was that Louise's obsession with David seems strangely unmotivated: She's merely a crazy woman who can't get over the guy who dumped her, even when she remarries (prefiguring later film characters, most famously the female lead Alex in *Fatal Attraction*). The original fact that the man was leading her on and continuing the affair—a much more rational cause for her emotional upset and confusion—was erased. Even if David remained a cad, without the adultery, Louise's disease became more delusional and unjustified, solely the result of her unbalanced feminine desire. (*Fatal Attraction* similarly reduced sympathetic elements for Alex during the production process.) The MPAA erased more legitimate causes for Louise's distress, further pathologizing her, when they insisted on an unhappy ending. A happy ending would have rewarded a criminal. Louise, even if mentally incompetent at the time, as the doctor clearly states, was nevertheless a murderer, and, as stated in more than one memo, the MPAA insisted they would only allow an ending suggestive of any possible happiness with Dean and recovery if it was made clear that this would take years. This became the film's ending: a clear retribution for Louise's moral failing, yet also inseparable from Louise's punishment for mental illness. The pathologized feminine was essentialized and medicalized, the former confirming the latter, rather than—despite the doctor's words—separating itself from it.

In *Possessed*, Louise appears simply as stereotypically feminine: overly sensitive and needy. The national board of censors erased the fact that a continuing affair was encouraging her. The ways a patriarchal, youth-centric society devalues aging, single, working women are never acknowledged as contributing to her anxiety. She merely wants, and not for particular and quite rational reasons (security, protection, finances, social approval, etc.). Similarly, the overwork and ambition of O. Henry's and other exhausted neurasthenics are rarely critiqued in terms of labor standards and expectations, capitalist expansion, or social upward mobility.

Pathological Technoculture

The pathologization, normalizing, and gendering of neurasthenia anticipate and resonate similar processes in pathological technoculture. I have taken the time for this extended discussion of examples of pathologization's normalizing and gendering work to prepare the reader for the complexity and at times contradictory nature of pathological technoculture. As seen with neurasthenia and schizophrenia, disease, ab/normality, class, gender, race, and other social categories are not stable. They are queer qualities of queer subjects, contradictory and flexible. Moreover, they are not things but processes—ongoing entanglements of phenomena, seen here in various material-discursive practices. The breadth of this chapter has been intended to demonstrate, not what disease is, but the scope of things disease can do. Such complexity and flexibility will recur in my later discussions of audiences and users, as well as the scientific and supernatural roots of technopathologies.

New disease science does not neutralize and displace old, inaccurate beliefs and diagnoses, but incorporates and imitates them, just as many media scholars have documented with new and old communication technologies: typewriter fonts and desktop visual metaphors appear on a computer, and vinyl LP crackle appears on digital recordings. Neurasthenia's patterns appear in schizophrenia and, a contemporary technopathology, electrosensitivity, which I will discuss later. There are even technopathological aspects to *Possessed*, as its much-remarked upon, new-at-the-time technique of cinematically recreating mentally ill subjectivity gave the viewer the experience of schizophrenia, through several first-person, (inter)subjective techniques meant to give insight into Louise's experience. One advertising campaign for the film suggested its ability to literally sicken the viewer. (Some editors had suggested the same about "The Yellow Wall-paper.")

Paranoid delusions of technological surveillance, manipulation, and torture are common symptoms of schizophrenia—all because of connection to unhealthy forces through electric communication technologies. Over the course of the twentieth century this focus of schizophrenic delusions had shifted generally from spirits, ghosts, and witches as sources of influence and persecution to radio waves, computers, and other electric communication technologies, especially television.[52] Neurasthenia was attributed to modern civilization, and in *Possessed* this link is reiterated in an opening scene. In the "Psychopathic Dept." of an L. A. public hospital, a young doctor bemoans the large number of mental cases they've been receiving: "One manic, three seniles, six alcoholics, and 10 schizos" in one day. The wise, older doctor notes that the rate is "going up all the time," and blames it on "this civilization of ours, [which] is a worse disease than heart trouble or tuberculosis. And we can't escape it."

Dr. Willard uses electrical metaphors when he speaks of Louise's condition, and what incapacitates can also cure. Static electricity and other electrical devices were a common treatment for neurasthenia; electroshock or electro-convulsive therapy (ECT) was and continues to be a treatment for schizophrenia.[53] Crawford was sued for invasion of privacy by a Pasadena mental patient whose ECT was observed, without her or her husband's consent, as part of the research for *Possessed*. The incident received a fair amount of press coverage before and during the film's release, although Crawford claimed to have done research only at LA General and not actually been present for the research trip to the Pasadena sanitarium. The degree to which ECT has been a cultural touchstone—it is not the most baroque treatment for mental health over the past century, yet it certainly is the most infamous—suggests the continued cultural salience of entanglements of electricity, biology, and mental health I will discuss in the cultural roots of technopathologies. Coincidentally, or perhaps contagiously, Pasadena was also where author Charlotte Perkins Gilman went to recover from her injurious attempted rest cure from neurasthenia with inventor Mitchell, described in "The Yellow Wall-paper." It was also the site of my focus groups for this book.

Doctor Willard's sober words to concerned husband Dean, at the film's conclusion, suggest pathological technoculture's persistent associations between electricity, biological energies, gender, technology, medicine, the supernatural, and disease.

> We human beings act according to certain patterns of behavior. Sometimes, why exactly we don't know yet, the pattern is broken—the *wires are crossed*. The mind cannot evaluate, judge, or even function properly. *Shock follows shock*, until eventually the mind gives way. The brain loses control and the body sinks into coma. Then, in a biblical sense, we might say that such a person is *possessed of devils*. And, it is the psychiatrist who must cast them out. (emphasis mine)

In addition to its techno-electrical metaphors, this dialog also connects *Possessed* to disease models more archaic than neurasthenia. "Devils" or evils spirits are one of the oldest etiologies of disease, and spirit possession was once thought to cause what we now consider mental illness. Moreover, the significance of this passage is evidenced by its prominent placement at the film's conclusion, its verbatim quotation in promotional articles for the film, and its singling out by the Pennsylvania Board of censors to be deleted.

Notes

1. Krajewski, "4 Housewives in 5 Suffer Ailment", 26
2. Gilman, *Difference and Pathology*, 23.
3. E.g., Douglas, *Purity*; Lupton, *Medicine*; Turner, *Body*.
4. Turner, *Body*, 176, emphases mine.
5. Bourdieu, *Distinction*; Lutz, "Curing the Blues," Veblen, "Conspicuous."
6. Anderson, "Mimickry," Bashford and Nugent, "Leprosy," Davis, "Constructing," Garland-Thomson, "Feminist Disability Studies," Tomso, "Queer History."

7. Browning, *Infectious Rhythm*; Douglas, *Purity*; Ewen and Ewen, *Typecasting*; Foucault, *Power*, 138; Gilman, *Difference and Pathology*; Lupton, *Medicine*; Turner, *Body*; Wailoo, *Drawing Blood*.
8. Gilman, *Difference and Pathology*, 23.
9. Metzl, *Protest Psychosis*.
10. "Ghetto Life."
11. Maines, *Technologies*.
12. Laqueur, *Making Sex*.
13. Gilman, *Difference and Pathology*, 24.
14. Beard, *Psychology*, Lutz, *American Nervousness*, "Curing the Blues"; Forth, "Moral Contagion," Gosling, *Before Freud*; Hoover, "Somatization Disorders"; Ng, "Culture-bound"; Schuster, *Neurasthenic Nation*; Schwartz, "Neurasthenia."
15. However, neurasthenia's flexibility led to its application in different countries at different times. Forth ("Moral Contagion") investigates its connection in fin-de-siècle France to a crisis in masculinity. Cowan, in *Cult of the Will*, investigates nervousness in Germany after 1890 as part of a cultural crisis in will power, expressive of the changes of modernity but also, he argues, laying cultural groundwork for fascism. Up until the 1990s, the diagnosis was a still used in Latvia (Skultans, "Historical Disorder") and China and among Chinese-Americans (Schwartz, "Neurasthenia").
16. Lutz, *American Nervousness*, 5.
17. Beard, *Psychology*, vi, emphasis original.
18. Ibid. 171.
19. Ibid. 189.
20. Prince, "Educational Treatment," 75.
21. Prince, "Association Neuroses."
22. Gilman, *Difference and Pathology*.
23. "Music and Neurotic Conditions in Young Girls," ¶1. Chlorosis was another gendered diagnosis, a type of anemia in young girls, abandoned in the twentieth century. See Wailoo, *Drawing Blood*, Chapter 1.
24. Beard, *Psychology*, 171.
25. Quoted in Lutz, "Curing the Blues," 140.
26. Quoted in Golden, *Charlotte Perkins*, 65.
27. Cowan, *Cult*.
28. McLaren, *Impotence*.
29. Lutz, *American Nervousness*, 310.
30. *Before Stonewall*.
31. Prince, "Educational Treatment," 100.
32. Gosling, *Before Freud*.
33. Lutz, "Curing the Blues."
34. Quoted in Prince, "Association Neuroses," 79.
35. Brock, *Health and Conduct*, 153.
36. Ibid. 261.
37. Brock, *Health and Conduct*, 22–23; 28.
38. Henry, *Adventures*, 30.

39. Ibid. 30.
40. Ibid. 37–38.
41. Gilman, "The Living," "Yellow Wall-paper," "Why I Wrote"; Golden, *Charlotte Perkins*; Heilman, "Debased Romanesque", "Overwriting Decadence"; Tuttle, "Rewriting."
42. Dumit, "'Come on'," "Illnesses."
43. Gilman, "Yellow Wall-paper," 144.
44. Gilman, "The Living," 49–50.
45. Sconce, *Haunted* and Milutis, *Ether*.
46. Gilman, "Yellow Wall-paper," 131, emphasis original.
47. Ibid. 134.
48. Ibid. 136.
49. Treichler, "Escaping the Sentence."
50. But not entirely abandoned: as recently as 2011, I had a friend tell me his mother had been diagnosed earlier in her life with neurasthenia.
51. "Inter-Office Communications"; *Possessed*: MPAA Production Code Administration File; Joan Crawford
52. Škodlar, Dernovšek, and Kocmur, "Psychopathology of Schizophrenia."
53. The neurasthenic notion of electrical treatment to manipulate gender performances continues to make appearances in popular culture. In Disney's two versions of *The Computer Wore Tennis Shoes*, an electrical shock / a lightning bolt turns a squeaky-clean / sub-average college boy into a mathematical genius / human encyclopedia. In the 1981 art-house pornographic film, *Nightdreams*, doctors use electric shocks to stimulate and observe a woman's erotic dreams. As Maines (*Orgasm*) describes, electricity powered the popular Victorian treatment for feminine hysteria, the vibrator, subsequent electrotherapeutics such as the violent wand also crossed over from medical device to sex toy.

References

Anderson, Warwick. "Going through the Motions: American Public Health and 'Colonial Mimicry'." *American Literary History*, vol. 14, no. 4, 2002, pp. 686–719.

Bashford, Alison, and Maria Nugent. "Leprosy and the Management of Race, Sexuality and Nation in Tropical Australia." In *Contagion: Epidemics, History and Culture from Smallpox to Anthrax*, edited by Alison Bashford and Claire Hooker, Pluto Press, 2002, pp. 106–28.

Beard, George Miller. *The Psychology of the Salem Witchcraft Excitement of 1692 and Its Practical Application to our Own Time*, G. P. Putnam's Sons, 1882.

Before Stonewall. Dirs. G. Schiller and R. Rosenberg. USA, 1984.

Bourdieu, Pierre. *Distinction: A Social Critique of the Judgment of Taste*, Harvard University Press, 1984.

Brock, Arthur J. *Health and Conduct*, Williams and Norgate, 1923.

Browning, B. *Infectious Rhythm: Metaphors of Contagion and the Spread of African Culture*, Routledge, 1998.

The Computer Wore Tennis Shoes. Dir. Robert Butler. 1969, USA.

The Computer Wore Tennis Shoes. Dir. Peyton Reed. 1995, USA.

Cowan, M. *Cult of the Will: Nervousness and German Modernity*, Pennsylvania State University Press, 2008.

Davis, Lennard J. "Constructing Normalcy." In *The Disability Studies Reader*, edited by Lennard Davis, Taylor & Francis, 1997, pp. 9–28.

Douglas, Mary. *Purity and Danger: An Analysis of Concepts of Pollution and Taboo*, Routledge, 2002.

Dumit, Joseph. "Illnesses You have to Fight to Get: Facts as Forces in Uncertain, Emergent Illnesses." *Social Science and Medicine* No. 62 (2006): 577–590.

Ewen, S., and E. Ewen. *Typecasting: On the Arts and Sciences of Human Inequality*, Revised ed., Seven Stories Press, 2008.

Forth, Christopher E. 2001. "Moral Contagion and the Will: The Crisis of Masculinity in Fin-de-Siècle France." In *Contagion: Epidemics, History and Culture from Smallpox to Anthrax*, edited by Alison Bashford and Claire Hooker, Pluto Press, 2002, pp. 61–75.

Foucault, Michel. 2001. *Power. The Essential Works of Foucault* 3, The New Press.

Garland-Thomson, Rosemarie. 2005. "Feminist Disability Studies." *Signs*, vol. 30, no. 2, pp. 1557–1587.

"Ghetto Life Cause of New Illness, MD Says." *National Enquirer*, 1969, December 14, p. 23.

Gilman, Charlotte Perkins. "The Living of Charlotte Perkins Gilman." In *Charlotte Perkins Gilman's The Yellow Wall-Paper: A Sourcebook and Critical Edition*, edited by C. J. Golden, Routledge, 2004.

Gilman, Charlotte Perkins. "The Yellow Wall-Paper." In *Charlotte Perkins Gilman's The Yellow Wall-Paper: A Sourcebook and Critical Edition*, edited by C. J. Golden, Routledge, 2004.

Gilman, Charlotte Perkins. "Why I Wrote 'The Yellow Wall-Paper'." In *Charlotte Perkins Gilman's The Yellow Wall-Paper: A Sourcebook and Critical Edition*, edited by C. J. Golden, Routledge, 2004.

Gilman, Sander L. *Difference and Pathology: Stereotypes of Sexuality, Race, and Madness*, Cornell University Press, 1985.

Golden, C. J., ed. *Charlotte Perkins Gilman's The Yellow Wall-Paper: A Sourcebook and Critical Edition*, Routledge, 2004.

Gosling, F. G. *Before Freud: Neurasthenia and the American Medical Community, 1870–1910*, University of Illinois Press, 1987.

Heilman, A. "'[A] Kind of "Debased Romanesque" with *Delerium Tremens*': Late-Victorian Wall Coverings and Charlotte Perkins Gilman's 'The Yellow Wall-Paper'." In *The Mixed Legacy of Charlotte Perkins Gilman*. C. Golden and J. S. Zangrando, eds. Newark, NJ: University of Delaware Press, 2000.

Heilman, A. "Overwriting Decadence: Charlotte Perkins Gilman, Oscar Wilde, and the Feminization of Art in 'The Yellow Wall-Paper.'" In *The Mixed Legacy of Charlotte Perkins Gilman*. C. Golden and J. S. Zangrando, eds. Newark, NJ: University of Delaware Press, 2000.

Henry, O. *Let me Feel your Pulse: Adventures in Neurasthenia*. New York: Doubleday Page and Co., 1910.

Hoover, C. R. "Somatization Disorders." In *Immigrant Women's Health*, edited by E. J. Kramer, S. Ivey, and Yu-Wen Ting, Jossey-Bass Publishers, 1999.

Krajewski, S. "4 Housewives in 5 Suffer Ailment Caused by Women's Lib." *National Enquirer*, 1983, March 15, p. 26.

Laqueur, Thomas. *Making Sex: Body and Gender from the Greeks to Freud*, Harvard University Press, 1990.

Lupton, Deborah. *Medicine as Culture: Illness, Disease and the Body*, Sage, 2012.

Lutz, Tom. *American Nervousness 1903: An Anecdotal History*, Cornell University Press, 1991.

Lutz, Tom. "Curing the Blues: W. E. B. Du Bois, Fashionable Diseases, and Degraded Music." *Black Music Research Journal*, vol. 11, no. 2, 1991, pp. 137–156.

Maines, Rachel. *The Technology of Orgasm: "Hysteria," the Vibrator, and Women's Sexual Satisfaction*, Johns Hopkins University Press, 1999.

McLaren, A. *Impotence: A Cultural History*, University of Chicago Press, 2007.

Metzl, J. *The Protest Psychosis: How Schizophrenia became a Black Disease*, Beacon Press, 2010.

Milutis, Joe. *Ether: The Nothing that Connects Everything*. Minneapolis: University of Minnesota Press, 2006.

"Music and Neurotic Conditions in Young Girls." *Medical News (1882–1905)*, vol. 74, no. 25, 1899, p. 793.

Ng, A. "Culture-bound Syndromes." In *Immigrant Women's Health*, edited by S. L. Ivey, E. J. Kramer, and Y.-W. Ting, Jossey-Bass Publishers, 1999.

Nightdreams. Dir. Francis Delia. 1981, USA.

Possessed. Dir. Curtis Bernhardt. 1947, USA.

Possessed. MPAA Production Code Administration file. Academy of Motion Picture Arts and Sciences Special Collections Library and Archive, Los Angeles, CA.

Prince, Morton. "Association Neuroses: A Study of Hysterical Joint Affections, Neurasthenia and Allied Forms of Neuro-mimesis." In *Psychotherapy and Multiple Personality: Selected Essays*, edited by J. Nathan G. Hale, Harvard University Press, 1975.

Prince, Morton. "The Educational Treatment of Neurasthenia and Certain Hysterical States." In *Psychotherapy and Multiple Personality: Selected Essays*, edited by J. Nathan G. Hale, Harvard University Press, 1975.

Schwartz, P. Y. "Neurasthenia." In *Immigrant Women's Health*, edited by S. L. Ivey, E. J. Kramer, and Y.-W. Ting, Jossey-Bass Publishers, 1999.

Schuster, David G. *Neurasthenic Nation: America's Search for Health, Happiness, and Comfort, 1869–1920*, Rutgers University Press, 2011.

Sconce, Jeffrey. *Haunted Media: Electronic Presence from Telegraphy to Television*. Durham, NC: Duke University Press, 2000.

Škodlar, Borut, M. Z. Dernovšek, and M. Kocmur. "Psychopathology of Schizophrenia in Ljubljana (Slovenia) From 1881 To 2000: Changes in the Content of Delusions in Schizophrenia Patients Related to Various Sociopolitical, Technical and Scientific Changes." *International Journal of Social Psychiatry*, vol. 54, no. 2, 2008, pp. 101–111. doi: 10.1177/0020764007083875

Skultans, V. "An Historical Disorder: Neurasthenia and the Testimony of Lives in Latvia." In *Colonialism and Psychiatry*, edited by D. Bhugra and R. Littlewood, Oxford University Press, 2001.

Tomso, Gregory. "The Queer History of Leprosy and Same-Sex Love." *American Literary History*, vol. 14, no. 4, 2002, pp. 747–775.
Treichler, Paula. "Escaping the Sentence: Diagnosis and Discourse in 'The Yellow Wallpaper'." *Tulsa Studies in Women's Literature*, vol. 3, nos. 1–2 (Spring–Autumn 1984), pp. 61–77.
Turner, Bryan S. *The Body and Society: Explorations in Social Theory* (Third ed.), Sage, 2008.
Tuttle, J. S. "Rewriting the West Cure: Charlotte Perkins Gilman, Owen Wister, and the Sexual Politics of Neurasthenia." In *The Mixed Legacy of Charlotte Perkins Gilman*. C. Golden and J. S. Zangrando, eds. Newark, NJ: University of Delaware Press, 2000.
Veblen, T. "Conspicuous Consumption." In *The Consumer Society Reader*, edited by J. B. Schor and D. B. Holt, The New Press, 2000.
Wailoo, Keith. *Drawing Blood: Technology and Disease Identity in Twentieth-century America*, The Johns Hopkins University Press, 1997.

· 2 ·

AUDIENCES AND USERS

A False Dichotomy of Entangled Subjects

In 1941, U.S. cryptographer William F. Friedman suffered a nervous breakdown. As discussed previously, this was one of the diagnoses into which the proto-technopathology neurasthenia evolved. Friedman's nervous exhaustion hit after he cracked Japan's "PURPLE" cipher machine—which used technology similar to a telephone exchange. "As always," writes German media theorist Friedrich Kittler, "it was precisely at the site of madness that machines originated."[1] Drawing on Kittler's work, Mara Mills notes that the frequent twentieth-century conceptualizations of communication technologies as prostheses or sensory extensions infer "that a generic 'disability' is somehow fundamental to the modern condition."[2] This situates technology and pathology at the heart of modernity.

Indeed, Kittler describes a shift in what he calls "discourse networks"—the technological systems for preserving and communicating knowledge and information—from nineteenth century romanticism to twentieth century modernism. Technologies of inscription, communication, and data storage are defining characteristics of the latter. Kittler calls *mediality* the technological conditions of possibility during a certain time within which literary forms and discourses of knowledge can take shape.

Pathology often arises in Kittler's descriptions of the mediality of modern discourse networks. Mysticism changes into hysteria, and reading is hallucinatory. The origins of communication technologies are rooted in blindness and deafness. Cinema implements mental mechanisms of madness. Headphones are unnatural. Machines take over functions of the central nervous system. The brain is like a phonograph.[3] In modern literature, women turn into celluloid and men into typewriters, and gramophones and radios influence turning women into prostitutes and men into poets.[4] All of these are due to an essential, technologically induced blindness.

> Circa 1900 several blindnesses—of the writer, of writing, of script—come together to guarantee an elementary blindness: the blind spot of the writing act. Instead of the play between Man the sign-setter and the writing surface, the philosopher as stylus and the tablet of Nature, there is the play between type and its Other, completely removed from subjects. Its name is inscription.[5]

Another way to think of this is a shift from producer to consumer. Instead of "Man the sign-setter and the writing surface"—the act of producing writing—being paradigmatic, it becomes the act of reading or consuming writing. Subjects encounter and consume type that has been removed from its act of inscription, the writer now a "completely removed" Other. In other words, the modern condition is one of audiences, not authors.

Kittler's ideas have been widely cited, and also critiqued. Media historian Lisa Gitelman, for one, aptly criticizes his tendency to naturalize or essentialize media, overstating changes in media history and under-citing his evidence.[6] However, I cite him here as an influential example of technology and media studies focusing on epochal historiography, describing ruptures and changes. As this is central to the mis-understanding of technopathologies as reactions to new media, it is important to unpack and challenge this tendency. Therefore, I turn to what is a related disjuncture or rupture presented in media history: the shift from passive, mass-media audiences who consume content to (inter)active new-media users who produce content. As with Kittler, pathologization is employed in differentiating two media paradigms, but I will argue that usership is not a new construct that displaces audiences, but one that developed alongside it.

Having previously described how pathologization categorizes on dimensions such as normality, gender, race, and others, I turn in this chapter to pathologization applied to technology users. Beginning with historical and etymological overviews of "audience" and "user" across discourses of media, technology, and health, I argue that pathologization has played a significant role in the evolving concept of "the user," but in ongoing entanglement with—rather than displacement of—"the audience." Audience and user too often have been placed in a false dichotomy and chronology, one that positions the interactive user displacing the passive audience. I will present my concept of users as tied to interactivity in communication technologies dating back to the telegraph and telephone. This is in contrast to, but simultaneous with, mass-media audiences. Rather than a disruption brought on by computerization, users have long coexisted with audience members in mutual constitution. Both, however, are constructs—and commodities. Just as neoliberalism is an acceleration of, rather than change from, modern capitalism, usership is a neoliberal expansion of audience-going, not a disjuncture from it. I draw on critiques of the celebratory hype of Web 2.0 interactivity by Jodi Dean and others to examine the user as an acceleration of audiences. I examine this process with a study of telephony as an important moment in interactive technology and the training of users, seen in instructional films for telephone use. As technopathologies start to appear during this period, we see how ill health is tied to individual technology use and users, not the generalized effects of mass-media content on audiences. Sickness individualizes; it differentiates atomized audience members into individuals, and puts responsibility on them. Users require education and instruction.

Audiences

The term "audience" derives from the fourteenth-century French noun with the same spelling, meaning "the action of hearing," as in when you are granted an audience with a sovereign. It is an opportunity to be heard. An audience is given or bestowed and, through this, an audience is also what is constituted among its recipients. This comes to refer to musical and theatrical performances, and, by the mid-nineteenth century, "audience" gains popular meanings that also include the readers of a book. This expansion of "audience" to mass-media communications continues into the twentieth century, as it applies to listeners of radio and experimental telephony, viewers of silent film, and listener-viewers of sound film, television, and their related offshoots, such as videotapes.

"Audience" is a singular, collective noun. There is no common term for a single audience member. The closest is a "spectator," which can be collected into a group of spectators. Although "spectator" carries connotations beyond my focus here—a primary association with sporting events, notions of the spectacular, and also the sense of being a witness, as to a car accident—it has been used extensively in cinema studies and art theory to theorize viewing relationships, power, and gender, among other aspects, at the individual psychological level. Given that my focus and method involve empirical evidence of discourse, rather than interior psychological states, I will not focus on spectatorship.

Audiences are the recipients of broadcasting, a term with roots in agriculture and religion. The agricultural origins have to do with debates around proper techniques for sowing seeds: planting in regular rows or "broadly casting" seeds about. As in the parable of the mustard seeds, the religious roots have to do with spreading the word of God. Both of these senses involve notions of growth, of cultivating crops and Christians. Indeed, religious tracts were the one of, if not the, first mass media. Rhetorically, Michael Warner defines an audience as persons who gather together to hear an argument. This framing in terms of persuasion is also about cultivation: the audience members will become either acceptors or rejecters of the argument.

It stands to reason, then, that concerns around mass-media broadcasting would involve what was being cultivated. Given that audiences were seen as part of a shared public sphere, they were involved in constituting national identity and building an "imagined community," and constructing national norms of race, ethnicity, and other identity categories.[7] There was great concern about audiences as cultures. I mean "culture" in the laboratory sense—a medium in which content is implanted and cultivated, like bacteria grown in a Petri dish. Intellectuals, such as those associated with The Frankfurt School, argued that the industries of mass media produced distracting spectacles with a homogenizing, pacifying effect on audiences, cultivating a mass culture. Pseudoindividualism and a rhetoric of uniqueness masked a pervasive and growing superficiality that numbed critical thinking, making citizens ripe for the groupthink of fascism and dictatorships. Similar concerns drove researchers with quite opposite ideologies and methodologies. Social scientists and psychologists investigated the effects of mass-media messages on individuals and communities. The emerging fields of mass-media research, propaganda studies, and media effects would form a cornerstone of the social science end of the current academic discipline of communication.[8] One of the more popular approaches, which adopts a longitudinal perspective of incremental effects through repeated exposure, goes by the name cultivation theory.

In mass communication research, and in the transdisciplinary research of media studies, two early examples of these concerns became foundational myths. Myths, not in the sense of being inherently false, but, as Mosco describes, narratives a culture retells to make sense of the world around them.[9] Two myths of mass-media audiences are the train-effect panic of silent film audiences and the *War of the Worlds* panic of early radio audiences. In one, early viewers of motion pictures were so astounded by the verisimilitude that they screamed and fainted at the projection of a train locomotive pulling into a station, fearing for their lives as if it were real. In the other, early radio audiences misunderstood a fictional play as actual news and believed Earth was being invaded by hostile Martians, some leaving their houses and hiding in the countryside. Both myths assert the fundamental naiveté of mass-media audiences and the power of mass-media content. However, scholarship has shown both to be exaggerations, which attests all the more to their mythic potency.[10] So, too, does the tenacious popularity of social-science media-effects research, which, after nearly a century, still offers largely mixed and inconclusive results.

By the late twentieth century, media studies had come to conceive of audiences as, not "cultural dopes," but significantly active in interpreting media content, although the degrees of freedom and constraint were debated.[11] However, even these models of the audience were criticized as too universalizing, in contrast to frameworks based on more concrete practices and structures of audience behavior.[12] As the media landscape changed with the rise of the Internet, mass communication studies were initially criticized for neglecting to apply their theories and research agendas to the emergent technologies.[13] Yet, at this stage, scholars still tended to reductively generalize "the" Internet as a singular medium. Moreover, the relevance of the audience as theoretical construct was seen as diminishing due to being too rooted in outdated mass communication theories, yet earlier considerations of audience-as-commodity labor were seen as increasing in relevance.[14] Various revisions were proposed, such as specifying and differentiating between audiences as mass, outcome, and agent.[15] Scholars acknowledged the variability of online audiences, such as the networked "imagined audiences" of social media accounts and producers.[16] Although the application of audience measurements to online users proved initially less than effective, the subsequent rise of online analytics, data scraping, sentiment analyses, individual consumer tracking, and aggregated consumer databases, are proving increasingly sophisticated and effective. The audience lives.

Users

In this book, I define users as persons who operate electric communication technologies to accomplish various tasks ranging from entertainment to business. While I do not argue that today's common definition of the term "user" exists transhistorically, the practices of usership do extend back over a century. Moreover, pathologization is a key differentiation between the two.

Whereas pathologization of audience members focused on fears of what would take root and grow from the seeds of media content, pathologization of users focuses, not on what they believe or become, but what they do, rightly and wrongly, with communication technologies.

Today, a "user" is commonly defined in opposition to audience members, conceptualized in individualized terms, reinforced by their actions of autonomously producing and/or uploading content, and therefore being more (inter) active. Online sources such as *Webpedia*, *Wikipedia*, *Dictionary.com*, and *YourDictionary.com* link the term "user" specifically to computational technologies: a user is someone who uses a computer. This rhetorical move positions users, and the interactive, networked, computerized technologies they use, as a break, a cut that establishes a difference from those who employed non-computerized technologies.

However, "users" are not new, nor is their defining hallmark of interactivity. As technology blogger Cynthia Rettig writes

> Let's face it, we are all "users." Anyone who accesses a computer software program has been known as a "user" for many years now. No doubt, the term originated somewhere back in the mainframe age when people had to sit at a terminal, log on, and identify themselves before they could access the mammoth machines, machines that filled whole buildings in the sixties.[17]

The *Oxford English Dictionary* dates the use of "user" in reference to a person who uses a computer system to a 1950 *Science* magazine article. However, the word "user" itself dates back to 1425, and it is first used in a medical context.[18] John Arderne, "the father of English surgery," uses the term in his famous *Treatises of fistula in ano: haemorrhoids, and clysters*. Arderne outlines the procedures of his greatest achievement, the treatment of anal fistula, which was an extremely dangerous medieval surgery. Arderne writes that it will be beneficial to the reader who follows the instructions and information provided: "It is not to doubt that if the foresaid be well kept that they shall give a gracious going to the user to the height of worship and winning."[19] This represents the *Oxford English Dictionary*'s primary definition of user, simply "A person who has or makes use of a thing, esp. regularly; a person who employs or practices something."

As with "broadcasting," the term also has a religious history, referring to one who uses God's grace and divine knowledge, as in the John Donne sermon from 1631, "As he [sc. God] sees him a good or a bad user of his graces." Indeed, today the notion of misuse is a defining connotation of the term "user." A "user" misuses the generosity of others, manipulates or takes advantage of them. A drug "user" misuses recreational or prescription medicine. An operating system or software application pleads in its own defense that a problem is due to "user error." However, a user who misuses technology bears responsibility for this, whereas audience members who cultivate bad messages generally do not, a point to which I will return. With audiences, content is the problem, requiring education and literacy to interpret content, and production of "better" content. With users, their own actions are the problem.

This can be seen in the term "end-user." Typically, it refers to the person for whom a software or hardware product is designed. The end user does not exist, but is a hypothetical, generalized ideal in the imagination of software and hardware designers. The end user is a perfectly discursive formation in that, by definition, it first exists during the design process of a new technology as a point of knowledge—who will be using this technology and what are their needs and habits—that precedes and thereby calls into material constitution the actual person who uses the technology. An end-user is imagined, a technology designed to fit them, and then actual users fit themselves to that technology as it has been designed.

This is an oversimplification, and there is a great body of research in social construction of technology, social shaping of technology, and related fields demonstrating how users reverse the direction of influence, exerting agency to change and reshape technologies. Indeed, these are bidirectional processes in which users and technologies co-construct one another.[20] Such research, while important, focuses on the material dimensions of users and technologies, which they show to be fluid to varying degrees.

However, my point, and the investigation of this book, is of "the user" as a subject position of discourse: our ideas and ideals about users and usership. The concept of "end user" in technology development provides an example of this. It is a technological version of the ultimate discursive subject, the consumer or the customer. Indeed, in economics, "end user" means simply consumer. *Investopedia* defines end user as "The true consumer of a product or service. The term 'end-user' is used to distinguish the person who will actually work with the good or service from individuals who are involved in other stages of its development, production and distribution."[21] Even in this broader context, the end user is a fully discursive construct, as the definition continues:

> In order to create a successful product or service, the people who create, develop, test and market it must think not of their own needs, but of the end-user's. For example, a computer programmer designing forex trading software would need to think about the level of computer knowledge and forex knowledge the end user will have (among other considerations) in order to design software that will accommodate the end-user's existing knowledge, ability and needs.[22]

The end user is a subject of knowledge. It is something *Investopedia* must "explain," and the subject of the thoughts of product creators, developers, testers, and marketers. Moreover, the specific subject of the discursive subject is discourse: What are the imagined end user's "knowledge, ability, and needs"? Finally, the end user is Other: product creators, developers, testers, and marketers must think "not of their own needs, but of the end-user's," which positions them as somehow different. This positioning of the (end) user as fundamentally Other or abnormal, at least to developers, foreshadows a point to which I will turn at the conclusion of this book: the user as normally abnormal, their pathology as standard deviance.

In the development of electrical computing, (end) users are positioned as a problem. For example, a 1983 field study of end users uses the term as an umbrella for non-professionals—common, emerging users of personal- and micro-computers who were neither computer scientists nor data processors by trade.[23] The growth of end-user computing is described as "one of the significant phenomena of the 1980s in the information management world," one which is defined as "interactive computing by non-DP (data processing) professionals."[24] The study identifies several problems this presents, such as users not practicing proper procedures for documentation, data storage, or security, as well as a fear that using a computer will "sidetrack" professionals from their existing career paths into becoming programmers, a low-status position at the time.

> One top manager stated, "Using a computer is essentially a clerical function. Executives and managers should tell their support people what they want from the computer." Many interviewees believed this to be a typical attitude in top management. … "Young managers don't want others to know they can use the computer." "I enjoy it, but I worry about what it will do to my career." "I don't want to become computer identified. It might lock me in a staff position."[25]

What is not explicit in the study, but suggested by terms such as "clerical," is that the risk here is gendered: masculine professionals being feminized by using what are perceived at the time as the equivalent of dictation and stenograph machines. Indeed, the original "computers" were typically women hired to do mathematical calculations, as shown in the film *Hidden Figures*. As discussed previously, feminization and pathologization are entangled. To try to address these concerns, the article recommends a series of educational and management strategies for dealing with the problem of end users.

During this period, the term "user-friendly" arises, but, even if the desired relationship is a friendly one, it still presumes a problem or difficulty. If an easy relationship with users were natural, there would be no need to consciously create interfaces that were "user friendly." User-friendliness again positions users as Other, as aliens, like a culture with different social codes. One cannot just *be* friendly to users, one has to anthropologically *learn* what friendliness means *to* users—which constitutes them as different from what designers and developers naturally understand as friendliness. Attempts to reverse this as user-unfriendliness in design can be seen, for example, in Joseph Corn's history of the concept, which extends beyond computers to automobiles and other technologies.[26] Although this shifts the "blame" or problem from users to designers, it maintains a division between users and designers as separate, distinct things with some sort of problematic gap dividing them.

In 1988, Stuart K. Card and Thomas P. Moran from the Xerox Palo Alto Research Center[27] offered the term "user technology" to outline a research agenda for creating workstations that afford more intimate and efficient human-computer interaction. Again, the user is a problem, an intellectual one. They describe part of the research must be uncovering "a technical understanding of the user himself"[28]—solving the problem of understanding the user. They trace their efforts to a program begun at PARC in 1974 to

> draw concepts from cognitive psychology and artificial intelligence to create an applied cognitive science of the user. … Computer science in general is extremely one-sided (for understandable reasons) in the treatment of its phenomena: almost no effort goes into understanding the nature of the human user.[29]

As part of this project, they define the user through four key interfaces: physical, cognitive, conceptual, and task. Users are problems on two levels: as a model for designers to create toward, and also a subject to be shaped by the software. "The challenge is to create systems that, through intimate cognitive interaction with users, aid them in structuring and manipulating their ideas … to get the user's mind around the [user's] ideas, as it were."[30]

In 1996, users are at the center of research in Human Computer Interaction, but still a problem to be solved: "User analysis is a crucial aspect of user-centered systems design," one study describes, "yet Human-Computer Interaction (HCI) has yet to formulate reliable and valid characterizations of users beyond gross distinctions based on task and experience."[31] Again, users are both a discursive model for designers and concrete persons to be influenced by the final product they created: "User analysis requires the design team to identify characteristics of the user population that are likely to influence their acceptance and effective use of an application."[32] The user is like early human anatomy—an arcane mystery needing to be dissected, made visible, communicated, and understood so that their functioning can be predicted, manipulated, and optimized.

Interactive Users: Not So New

Although conceptions of audience members and users have fundamental differences, users are not new. Audiences and users have been in mutual counter-distinction, entanglement, and constitution throughout the history of electric communication technologies. Technology historian Sue Currell, for example, describes constitution of users in training technologies for speed-reading in the late nineteenth and early twentieth centuries, a perspective amenable with Jonathan Crary's work on the nineteenth-century evolution of observers, in which he differentiates spectators from observers, who not only observe visible phenomena and media but also rules, practices, and procedures—a process of subjectification that sounds much like usership—and Jonathan Sterne's history of cultural practices and conceptions of mediated listening.[33]

What about interactivity? Interactivity has been cited as the defining characteristic of "Web 2.0." As the conventional narrative goes, in the switch from passively viewing web pages, new technologies enabled the audiences of websites to actively respond to, manipulate, and easily create their own web content. They evolved into users of networked communication technologies. This narrative hinges on the rhetorical work of the appellation "Web 2.0" in suggesting a dramatic change, and it was celebrated as not only a technological breakthrough in interactivity but also an emergent democratic social movement.[34] The false dichotomy of old audience member/new user posits that interactivity is a defining difference, a key sign of technological progress that elevates passive audience members into new realms of agency as technology users, with broad social implications.

However, interaction with media has at least a century of history, including nonelectric media, such as print.[35] More recent examples include the 1950s children's television show *Winky Dink and You*, the branching narratives of *Choose Your Own Adventure* novels, the 1967 Czech film *Kino-Automat*,[36] QUBE cable television, audience-as-jury plays *The Night of January 16th* or *The Mystery of Edwin Drood*, and multi-angle-switching DVDs. User-interface historian Anker Helms Jørgenson suggests that "interaction" came to replace terminology in 1950s office automation studies and 1960s–1970s user-interface computing research that was more explicitly communicative and less individualistic: *language*, *dialects*, *understand*, *talk*, *conversational*, and *dialogue*.[37] Use of the term *audience-participation* to describe interaction is first recorded in 1940. Lisa Gitelman has shown how roles of producer and consumer were already blurred during the emergent period of the phonograph.

Indeed, audiences have always interacted: they choose to attend and listen, or to leave or ignore. They sigh, whisper, and shift in their seats. They applaud or withhold. They purchase tickets, or do not. They change channels. They go to the bathroom during commercials. They write letters of complaint to production and distribution companies. They actively make meaning, contesting, reshaping, and negotiating or "decoding" the manifest content in media messages. They create fan 'zines, slash fiction, mix tapes, and home-video parodies, and exert other forms of creative labor that enrich and redefine their viewing and listening experiences. Interactivity is neither new nor the sole terrain of computer users; they merely represent an intensification of it.

Paul Virilio argues that "Interactivity was actually born in the nineteenth century—with the telegraph, certainly, but also and especially with clinical electricity, which involved planting electrodes on the faces of the human guinea pigs used in such 'medical art' as practiced by [La Salpetriere's] Dr. Duchenne de Boulogne."[38] His assertion connects usership to medical science: The use of electricity to interact with the medium of a patient's body, to manipulate it actively rather than be a passive prisoner to it, evokes notions of today's interactive user. This exploration will be discussed further in the next chapter, but, for now, as noted previously, "user" has medical etymologies.

Neoliberalism

Audiences are vulnerable populations to be protected—from propaganda, dumbing-down, provocation, and panics—but users are individual problems to be solved. I do not see usership as a rupture from audience members, but an intensification and exaggeration of them, one in which social concerns transform into individual responsibility. As such, the user as a construct fits within theorizations of neoliberal economies and subjectivities.

David Harvey describes neoliberalism as both "a utopian project to realize a theoretical design for the reorganization of international capitalism [and] a political project to re-establish the conditions for capital accumulation and to restore the power of economic elites."[39] He finds the political project to have been successful, whereas the utopian project is more a theory used to justify and rationalize the political project when needed. This theory argues "that human well-being can best be advanced by liberating individual entrepreneurial freedoms and skills within an institutional framework characterized by strong private property rights, free markets, and free trade."[40]

Harvey traces the neoliberal turn to the Great Depression of the 1930s, which prompted a postwar preventative effort. In the 1950s, think tanks and social scientists promoted and theorized neoliberal positions. The inflation and unemployment of the 1970s provided motivation to implement neoliberal policies, particularly under the subsequent administrations of Thatcher in the UK and Reagan/Bush in the US, and continuing since then.

Neoliberal society expresses the cultural values associated with these political changes, and neoliberal subjectivity the experience of living within them, the ideals and norms they set up. First and foremost is ruthless individualism, a priority on individual (rather than state) responsibility, problem-solving, and entrepreneurship (rather than a lifelong employer and corporate pension). Individual responsibility (as persons or representatives of organizations) is prioritized through an emphasis on contractual relations. Rhetoric that triumphs the power of individual freedom, equality, and resultant prosperity, sounds very similar to the rhetoric of interactive or postfeminist empowerment.[41]

Indeed, the timeline of neoliberalism roughly reflects the development of electric communications technologies. Early experiments and deployments occur up through the 1930s. Postwar, the transition to mass-media, audience-based broadcasting is largely complete. However, users do not appear out of nowhere, but develop steadily until the mass adoption of computing technologies, roughly parallel with the mass adoption and implementation of neoliberal politics and practices. Technologies are key, for, as Harvey writes, neoliberalism "requires technologies of information creation and capacities to accumulate, store, transfer, analyze, and use massive databases to guide decisions in the global marketplace. Hence neoliberalism's intense interest in and pursuit of information technologies."[42] Importantly, as Manuel Castells points out, information and communication technologies do not cause neoliberalism, or his broader theorization of "network society," but are components of already occurring social trends and changes.

The rise of the construct of the interactive user is a progress narrative of the triumph of the new. Progress narratives of technological change derive from and can be traced through other deep-rooted teleologies, such as American exceptionalism,[43] Enlightenment science, and Western redemptive spirituality.[44] As part of unpacking such ideological origins, several media historians have undertaken demystification of the "new" in new media.[45] Bermejo reminds us that audiences are not natural phenomena or theoretical constructs but, within commercial media systems, manufactured commodities produced by advertising-supported media, developed at the beginning of the twentieth century as a way to manage mass consumption. As commodities, audiences also labor in seeking out and engaging with media content (or the reverse), activity which must be measured and correlated with consumption. Ultimately, Bermejo questions whether the "user" is simply a new replacement for the commodity "audience," or if more complex concepts are needed.[46]

From an economic perspective, mass audiences can be understood as a product of modern industrialization and Fordist standardization. The associated cultural concern is the capitalist destruction of individualism, and the replication of individuals as standardized as widgets on the factory lines where they work. They passively bear the stamp of the industrial imprint. Users are a neoliberal, post-Fordist framework. For example, Briziarelli's ethnography of Italian Facebook users argues that the social media platform harnesses ideologies of free labor to shape neoliberal subjectivity.[47] Gershon sees a similarly neoliberal logic induced by Facebook being applied into interpersonal relationships.[48] These represent new, dynamic, reflexive subjects of fluid economies, engaged in projects of self-creation and adapting to constantly changing labor expectations. They actively create themselves within flows of a networked, global economy. Usership here can be understood as a contract between the individual user and the benefits of the technology. In line with neoliberal individualism, the user bears a heightened responsibility for their proper and correct use, particularly in how it affects others. Users, as problems to be solved, require training. Such training evidences their potential for misuse and abuse.

Users, like neoliberalism, are not new, but an acceleration of audiences. As Harvey describes, neoliberalism is not the opposite of modern capitalism. It does not represent a historic closure but an exaggeration or intensification.

In *The Social Transformation of American Medicine*, Paul Starr describes a long, two-pronged process of the increasing sovereignty of medical professionals and the corporatization of the medical profession. A key part of these movements, from the postwar period to the 1980s, involved moving medical care out of the home and into the marketplace—a strategy of commodification that clearly represents neoliberalism as intensified capitalism. Similarly, I argue that the user, contrary to many new-media claims, is not an opposing disruption of a new category, but an intensification and exaggeration of the pre-existing category of audience members. Hayward, for example, traces early formations of neoliberal subjectivity in technologies of the teleprompter, photo-booth self-portraits, and automated teller machines.[49]

The myth of a rupture in interactive usership with far-reaching implications—aka "Web 2.0"—has economic origins, specifically in marketing. Business-to-business marketers CMP Media trademarked "Web 2.0" in 2003.[50] In 2004, information technology tradeshow producer Medialive International used Web 2.0 as title of a business conference.[51] Their partner was technology publisher O'Reilly Media, "an active participant in the technology community ... [with] a long history of advocacy, meme-making, and evangelism".[52] Here, the term was intended to refer to the conference theme of "the Web as platform." This did not suggest a major new release, update, or version of the Internet or World Wide Web, as the numbering convention from software releases typically suggested at the time. Those who appropriated this numbering convention were no doubt aware of its dominant meaning and use. However, Web 2.0 was shifted from a numeric convention for identifying significant (i.e., you must buy) software upgrades to their use.

Web 2.0 now referred to not merely new capabilities built into an application, but to the new ways *people* were using the Web to distribute, collaborate, reconfigure, and share information. The suggestion was true interactive communication and collaboration, rather than mere reading and publishing discrete pages and sites. An article in *Digital Web Magazine* described this as a move from static documents to more supple data. The users of new "microcontent"—harnessing RSS, Flickr, Delicious, semantic markup languages, Google Maps, podcasting, etc.—could share, remix, deconstruct, and recombine rather than merely read or write. The "new" element of user control was touted as a "paradigm shift" that would affect "the people who use it socially, culturally, and even politically."[53]

Yet, the marketing connection was still prevalent: "Because content flows across the Web in RSS feeds and can be remixed along the way, Web designers must now think beyond sites and figure out how to *brand the content itself*".[54]

The social, user-focus of Web 2.0 steadily grew in recognition. In 2005, the religious tract of techno-evangelism, *Wired* magazine, proclaimed, "We are the Web … behold the power of the people"[55] By the end of the following year, *Time* proclaimed "you"—as exemplified by Web 2.0 poster child YouTube—its Person of the Year (2006–2007). CNN venerated Web 2.0 social networking with its headline, "Will the 2008 USA election be won on Facebook?"[56]

Several scholars of new media and technology have proposed more skeptical approaches to the celebratory, empowering claims associated with participatory technologies, such as those of Web 2.0, asserting that technologies' cooptation, inherent design, or unintended effects perpetuate rather than challenge existing structures of power, media, and politics, reinforcing neoliberalism, globalization, surveillance, self-enterprise, and profit-seeking.[57] Moreover, the construct of interactivity has been critiqued for generalizing new media and reducing uploading to a linear process.[58]

In particular, Jodi Dean's theory of "communicative capitalism" focuses on interactivity from an economic and social perspective.[59] Within communicative capitalism, the privileging of individualism, victimization, and consumerism undermine progressive politics and evacuate the term "democracy" of meaning. Networked communication technologies are key to this process, as users participate in a frenzy of interactive posting, messaging, and liking, which transform the necessary debate and conflict of true democracy to empty circulations of feelings and opinions. This serves to obscure the Internet's role as infrastructure for global capitalism and maintain fantasies of interactive, democratic participation with meaningful effects.

The rhetoric of Web 2.0 argued that technological interaction spawned or reflected increasing social agency and equality. However, one can reverse this equation: Web 2.0 served to rhetorically imbue using new functionalities of primarily entertainment media commodities with individual empowerment, social change, and a rush of meaningfulness. It harnessed the effect of democratic soldiering to the equivalent of a remote control or picture-in-picture display. For example, consider an early, popular narrative of tech-enthused, populist participation and democratic empowerment, the mobile-messaging-organized crowds of protesters who brought down Philippine President Estrada, an event celebrated by techno-enthusiasts as an exemplary "smart mob."[60] Vincente Rafael complicates this with cultural and historic specificity, skepticism toward its generalizability, and tempered celebration: Estrada's tech-fueled, bloodless coup merely replaced him with his vice president. Significantly, the name of the techno-social movement that brought about these changes—"Generation Txt"—was not born organically by comrades around a campfire, but originated in a marketing campaign.[61]

However, only a few years later, social media—the latest incarnation of Web 2.0—were lauded for their role in democratic protests in the Middle East. The "Arab Spring" was variously referred to as a "Twitter Revolution" or "Facebook Revolution," spawning popular and scholarly debates about the role of these technologies in social and political change. Arguably, this is communicative capitalism at work: a focus on circulation of messages and an exhilarating fantasy of the power of participation that distract from, not only the work of actual politics, but the expanding role of the Internet as the infrastructure of global capitalism and its inequalities. Russian use of social media to interfere in the 2016 US elections provides stark contrast to ealier narratives of interactive technologies inherently aiding democracy.

Usership is not new. Nor is it different from audience membership; it is an advanced, exaggerated, neoliberal form of it. Interactivity is not a marker that differentiates audiences from users, it is a vector of exaggeration and acceleration by which usership is differentiated from audience membership. Marx describes the capitalistic process of the body re-shaping itself to fit the form and needs of the tool (instead of the tool serving the user), as a form of possession. The tool possesses the user. Foucault theorizes the possessed body as, not a battle lost, not a body of submission, but as a site of struggle between opposing forces.[62] The physical convulsions of the possessed demonstrate the struggle taking place. Technopathologies present bodies in struggle, entangled processes of modern capitalism and its advanced form of neoliberalism. For example, globalization, as a component of neoliberalism, accelerates hatred of the diseased and infirm, among other minorities, as Appadurai argues they serve as reminders of the failures of nationalist projects of purity and wholeness.[63] Rather than opposing forces, these are levels of degree. The struggle is a transition, a birthing process of being pulled in multiple directions, which creates the convulsion, the sickness, the struggle.

Pathology is a dimension of that process. Pathologization defines usership by delineating the limit cases of interactivity. If a user can do things, they can do the wrong things, or do things wrongly. Usership can go wrong, can go bad, can be sickening. Pathology is a chiaroscuro that carves out the figure of the user.

Users in Training: Early Telephony

Users require education. Audiences might need literacy skills in interpreting content, but not in *using* media. The ease of use of radio, TV, and cinema for audiences has long been held up as an example of what computerized technologies should aspire to—"plug and play" simplicity and usability—yet rarely achieve. Indeed, "usability studies" are rarely conceived of in terms of broadcast technologies. In contrast, a tool requires proper use. There are proper and improper ways to use, good and bad users. One dimension of demarcating this is disease. Sickness makes visible incorrect use, bad users. The prevention of sickness is education.

Telegraphy hints at early usership in public displays of its use. Plotnick describes, in the open in ceremonies of world fairs, the spectacle of U.S. presidents using telegraphy to initiate the fairs through "long-distance button events."[64] Widely covered in the media between 1884 and 1933, Plotnick situates these events within the "historical moment when the introduction of electricity both reinforced and reconstituted how individuals physically interacted with technical objects."[65] Such physical interaction with technologies is the essence of usership, a novelty at the time worthy of spectacle. A 1909 French Naval surgeon researched whether wireless telegraphy was causing an increase in neurasthenia among naval men. Although dismissing the concerns as outdated, he stopped shy of entirely dismissing the association: "The report just received from New Jersey of the death of six women in a wireless testing department from nervous collapse followed by fever and unconsciousness will therefore not be debited to the action of long wave currents until exact information is received."[66]

Telegraphy, however, required the intermediary of the telegraph operator to use. Telephony presented a transition: it both required an intermediary operator and had "plug and play" simplicity in direct use by the subscriber.

As several scholars have noted and Mara Mills has explicitly traced, pathology was, from an early time, part of the mutual constitution of technology and users in telephony, with areas such as deaf education, assistive technologies, speech wave visualization, phonetics, and communication technology development engaged in mutual influence for decades.[67] During the transitional period that I am examining, Mills notes that American Telephone & Telegraph had started a major study of speech and hearing as early as 1913. In the 1920s, the company collaborated with a group working for the establishment of "hard of hearing" as a diagnosis distinct from "deafness," and collected the first large-scale data on auditory thresholds as well as developed the first commercial electronic audiometer. Mills writes, "their collaboration had the unintended consequence of narrowing cultural standards for 'good' communication. … Telephone engineers adopted the concept of 'deafening,' abstracting it from physiological injury to include temporary 'hearing impairments' caused at any point along a telephone circuit: room, transmitter, line, repeater."[68] In other words, defining and differentiating pathological categories were directly entangled and mapped onto technological categories of ideal or normal sound (in contrast to interference or noise) and, thereby, normal hearing. As a result of this, Mills writes, "Hearing loss became a communicable disease in the early twentieth century,"[69] contagion being a concept to which I will turn in a subsequent chapter.

During this period, successive iterations of telephonic technology were rolled out at different rates across the United States. While other communication technologies also participated in linking the national expanse, creating a mediated public sphere, and disciplining subjects, telephony was unique in several qualities. It was immediate (unlike print or film), two-way (unlike radio's eventual development), and largely operated and experienced directly by its users, with the exception of operator assistance (unlike telegraphy). Comparatively, telephone users had more agency and participation in this national electronic medium, making their protocols of usership more necessary in establishing proper technological citizens, particularly given a medium primarily of sound, which has been theorized as especially relevant for the constitution of subjectivities.

Telephone users were instructed in many ways. A 1917 interior decoration book calls the telephone an "abomination" that should be hidden upstairs or in a closet.[70] In its early decades, educational and instructional guides insisted on using telephones for practical, business matters, not socializing, replicating the economy of the price-per-word telegraph, until social calls were recognized as profitable in the 1920s. User abuse of shared party lines was depicted for decades, everywhere from Hank Williams' country song "Mind Your Own Business" to the Rock Hudson/Doris Day romantic comedy film *Pillow Talk*. Etiquette and courtesy books in the 1900s–1920s also gave advice for how subscribers/users should treat operators.[71] Advertisements instructed rural women how telephone use could solve their loneliness (Fig. 2.1).

Indeed, the operator was now a partner with and hence part of the user, and we see a focus on being a proper operator as well. It has been argued that the blended subjectivities of the operator and user partially caused telephony's poor early adoption in France, where it went against the cultural grain of centralized order and hierarchy, and was actively discouraged by the government.[72] The most famous example of educating and disciplining American operators was the replacement of the original telephone boys with young women, who

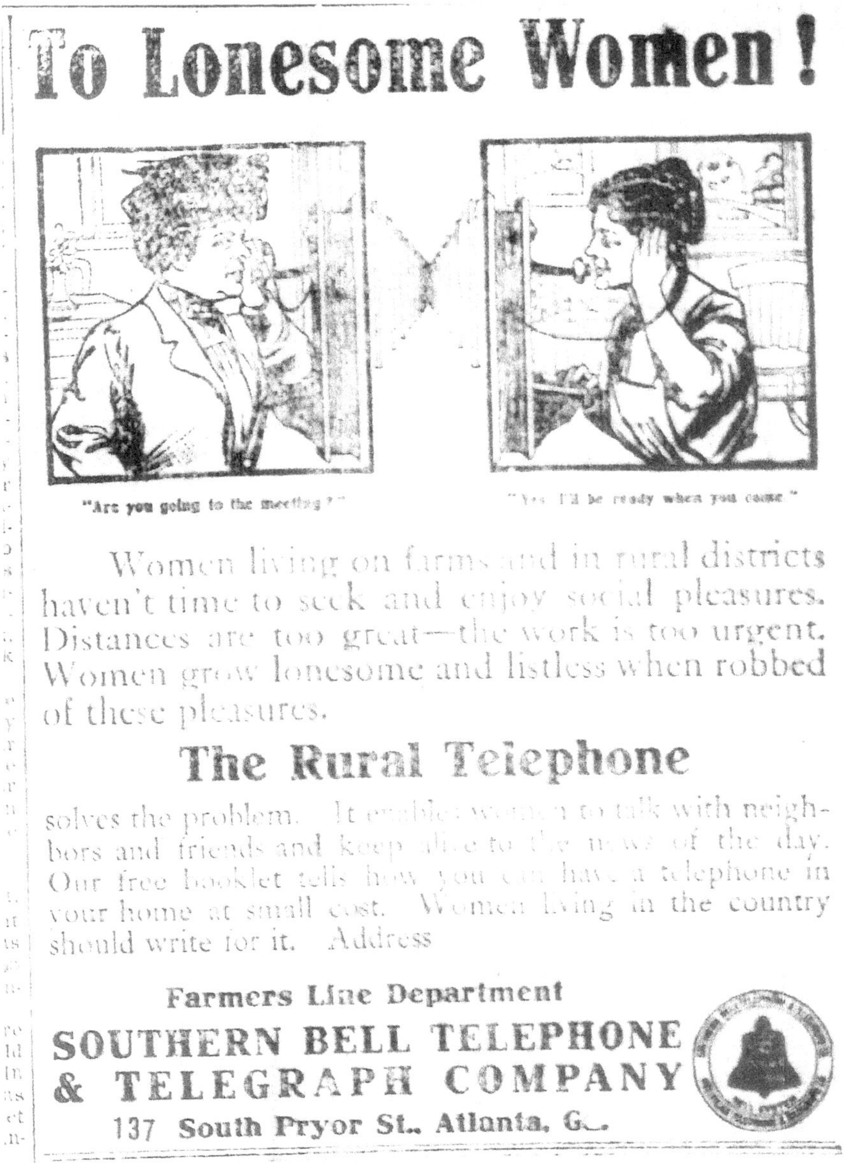

Figure 2.1. "To Lonesome Women!" Rural women instructed on telephone use as cure for loneliness. Advertisement for Southern Bell Telephone.
Source: *Anderson Peoples Advocate*, 20, 93 (1910, Sept 29): 3.

were deemed more patient, polite, subservient, and affordable than their rowdy, impolite predecessors. However, this required ongoing discourses of proper and improper usership. Female operators were frequently depicted using feminine stereotypes of stunted intelligence, arrested development, nosiness, and general ineptitude, and compared to lunatics and imbeciles.[73] Operators were satirized in musicals such as *Bells are Ringing*, *Auntie Mame*, and *The Telephone Girl*, as well as a 1927 film comedy by the same name.[74] Later in the century, comedian Lily Tomlin's Ernestine character personified the obnoxious operator-as-bitch, and even mechanized operators—answering machines—were blamed as part of the "entire problem" of telephone rudeness and lost manners in a letter to the *National Enquirer*.[75] As recently as 2000, in the film *The Operator*, a femme-fatale telephone operator who calls herself "Shiva" wreaks havoc on male victims.

A number of short films were made to instruct telephone operators and users. *Dial Comes to Town*, *Now You're Talking*, and *How to Use the Dial Phone* are three of numerous telephone training films.[76] In them, we see people instructed and educated in how to use the phone—how to be proper users. These are guides to proper interactivity. The audience constituted by these films is not an audience, but users. In these films, the distinction and constitution of ideal users is entangled with existing social hierarchies. In mapping the contours of good and bad users, telephone training films also harness social stereotypes pertaining to gender, age, and race, imbuing seemingly neutral technological practices with hierarchical power relations of different social categories.

For example, *How to Use the Dial Phone* and *Dial Comes to Town* associate stereotypical feminine clumsiness and predilection for gossip with improper telephone usage. *Now You're Talking* also deploys racial stereotyping in its depiction of an African American zookeeper (Fig. 2.2). Mostly denied voice, he mistakes a wrong-number call for "Mr. Lyons" as a call for the lion, to whom he serves tea, only to be attacked. In title cards, the caller says, "Yaas, I want to speak to Mr. Lyons." The zookeeper replies, "Ah was feeding him, Sah, and he bit me!"

In *Dial Comes to Town*, age is associated with abnormal usership. "Pshaw!" mutters Grandpa, who is indifferent, if not directly hostile, to the arrival in his unspecified rural U.S. community of a new telephone technology. Telephone users will soon input numbers directly, and dial tones will replace operators. Mother and Father smile, shaking their heads at Grandpa's foolish resistance to progress. His perky young grand-daughter looks up from her homework to proclaim that she thinks modern telephones are "Yummy!" As younger characters explain, Grandpa's dismissive reaction to the new telephony parallels his previous encounter with a new washing machine. *Dial Comes to Town*

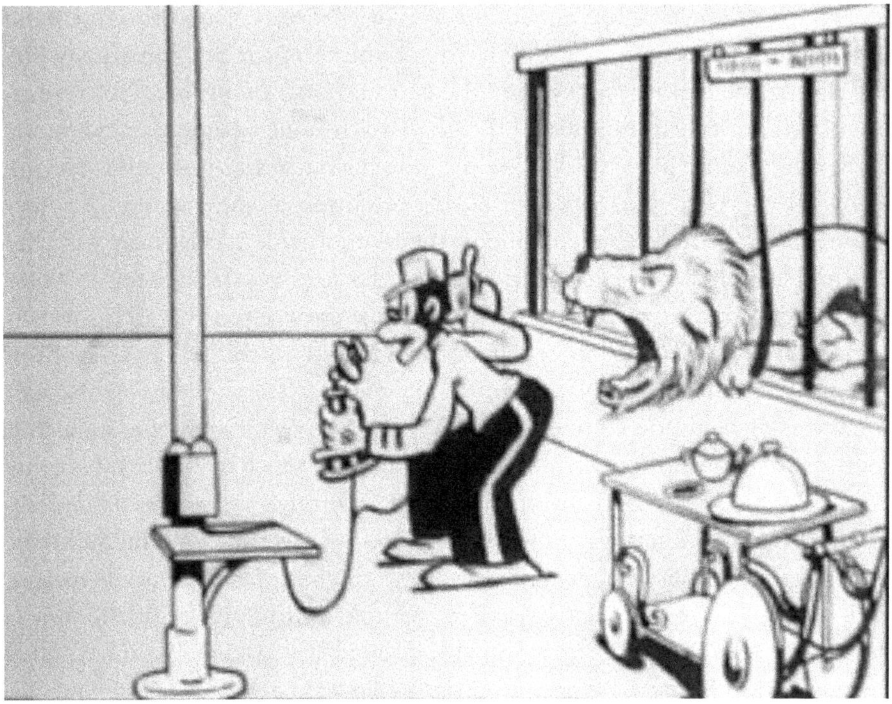

Figure 2.2. Racist stereotypes used in training telephone users.
Source: From *Now You're Talking*, directed by Dave Fleischer, scenario by Max Fleischer (Inkwell Studios, 1927). Screen capture. Internet Archive. Public domain.

pathologizes elderly and nonurban users as irrationally resistant to technological progress and specifies an inappropriate user of a party line as an older, matronly woman, who receives dirty looks for her eavesdropping at the town meeting. The film depicts what Fischer notes: that conflicts over appropriate phone usage were most commonly associated with rural users, and that female users were most associated with trivial conversations.[77]

Indeed, in *How to Use the Dial Phone*, it is a young, attractive woman who must learn appropriate usage through trial and error. During two examples of what not to do, a strange, white, cardboard cutout of a male hand appears in the frame of an otherwise realistic, live-action film. A title card reads, "You will get the wrong number IF—Dialing is started before taking the receiver off the hook." A close-up of the phone directory and phone follows, then a more extreme close-up of the female protagonist's index finger correctly engaging the rotary dial for five digits (at the edge of the frame the receiver, unmolested). In the next shot, however, we see the bad example: a medium, over-the-shoulder shot of the woman dialing features the white hand in the left of the screen, pointing to the phone receiver. She incorrectly lifts this during the dialing process, and the pointer hand disappears. This repeats in a subsequent demonstration: "IF—the switch hook is moved accidentally during the dialing process." We see the young woman full-front now, but she seems busy, preoccupied, and not paying attention as she reaches for something on her writing table and bumps the switch hook accidentally with her arm and loses her call. Having to hang up, she frowns. The masculine white hand directs our attention to the switch hook the entire time. (This shot is either the film's conclusion or where the currently archived copy has been severed.) The hand literally illustrates male authority in guiding us through the errors of feminine technological use and, as a communicative indicator, suggests male authority in the feminist-Lacanian sense of the Symbolic-as-patriarchy.[78] As we watch a woman stumbling through inappropriate dialing, in an instructional narrative on being a good user, we see a literal and symbolic representation of male authority enter the film, penetrating its realism with symbolic, communicative artifice. Moreover, the viewer of the film is positioned as a good user by being able to read this: literate in the English language and viewing conventions of early silent cinema, possessing symbolic competency in knowing what to make of a strange white drawing of a hand poking into the frame, and able to coherently integrate all three. Such positioning of the viewer as a good user pathologizes the female object of the film all the more through contrast and differentiation as a bad user.

From Education to Pathologization: Emergent Technopathologies

It is a short progression from improper use to unhealthy or otherwise dangerous use, and technopathologies are discourses that define abnormal use and thereby, users. Electric communication technologies could make you sick. Early studies of telephone switchboard operators found a high incidence of nervous illnesses.[79] Early telephone users experienced stage fright and "telephone ear"[80] and, if overusing the technology, "aural overpressure," a condition of recurring ring echoes, excited nerves, tinnitus, and "giddiness and neuralgic pains."[81] Berlin telephone operators in the 1920s suffered nervous collapses from the electric shocks they received while connecting hundreds of calls an hour—but were dismissed as just wanting attention.[82] Around the same time, early radio operators experienced "listeneritis" and "radioitis," the hallucinatory "phenomenon of the radio operator hearing something in the static where there was nothing at all."[83] In the early years of sound movies, the unnatural combination of single-perspective visuals with multi-perspectival audio caused confusion and disorientation.[84]

Technologies could also exhibit another type of disease relationship, that of a carrier or conduit. In the 1890s, users worried about putting a telephone in the room with a sick person, lest calls transmit illness to the other caller.[85] The journal *Medical News* reported in 1899:

> President Murphy of the New York City Board of Health, believing that telephone transmitters are a means of spreading disease, has ordered inspectors to examine the public telephones. The inspectors will thoroughly cleanse the transmitters and receivers with cotton, which will then be sent to the Bacteriological Department of the Board of Health. Should it be found that telephones are likely to disseminate disease organisms; an order will be issued requiring daily disinfection of the instruments by the telephone company.[86]

In 1909 The Baltimore Electrical Supply Company offered for sale a "sanitary transmitter mouthpiece," consisting of a brass clamp and glass or porcelain mouthpiece that could be removed for sanitation (Fig. 2.3).[87]

Public phone offices, early places for shared telephone use, provoked public health concerns as contagious locales. In the days of early silent cinema, when nickelodeons and arcades were more the norm than stately movie palaces, there was a public health panic around the mixing of classes in these environments. It was not merely that upper-class patrons might be discomfited by mixing with the lower classes while enjoying movies, but also that the poor ventilation and enclosed spaces were seen as increasing the likelihood of carrying one of the many infectious diseases borne by the lower classes.[88]

Telephones, their wires, and poles were feared as transmitters of not only germs, but also dangerous lightning and electrical shocks.[89] As a newspaper in central South Carolina described in 1913:

> It is well known that steel is a conductor of lightning, and telephone and telegraph wires most frequently attract the bolts that cause poles and other objects to be shattered, and yet people are so engrossed with business affairs, or careless as to observing ordinary care, that they will use a telephone during severe storms.
>
> This is very dangerous, and should be avoided, not only for the sake of the person using the telephone, but for the central operators as well. Every year there are recorded deaths because people use telephones during storms, and not only last

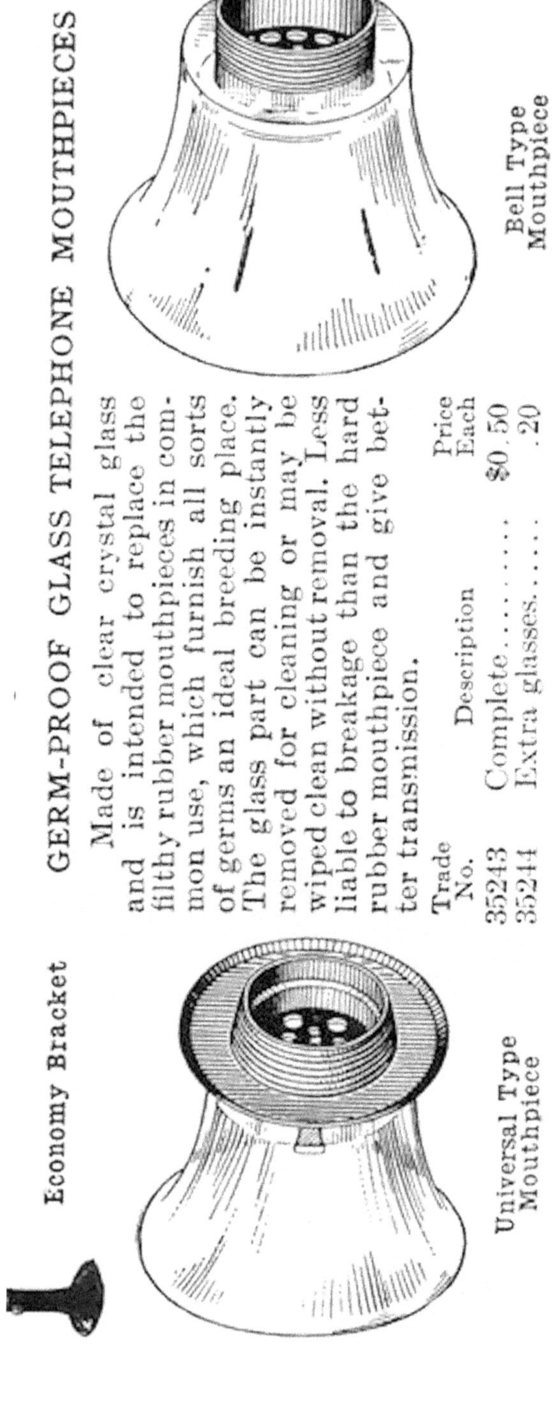

Figure 2.3. "Germ-proof Glass Telephone Mouthpieces." Devices to prevent telephones from being carriers of disease.

Source: *The Baltimore Electrical Supply Company* commercial catalog, 1909. Huntington Library and Archives. Electrical lighting catalog collection, Box 6, item 93, p. 945.

> summer we knew a young man who was knocked unconscious and remained dazed for hours because he answered a telephone that was rung during a bad storm.[90]

In such a description, we see the user clearly described as facing a health risk from technology misuse and, in neoliberal fashion, it is their own responsibility. Moreover, as I will discuss in a later chapter, they not only threaten themselves but others.

The telephone continued to be a locus of pathological discourses. Even though, by 1903, the United States had three million telephones and led the world in telephone density by 1914, in 1971, one study of journalists included reports of ear pain and heat from extensive telephone use; another documented dysesthesia of the scalp from mobile phone use.[91] In 1993, Dr. Jacob D. Rozbruch, chief of orthopedic surgery at New York's Beth Israel Medical Center North, coined "Free Hand Syndrome" to describe nerve injuries and chronic pain caused by cradling a telephone receiver between head and one shoulder in order to keep hands free.[92] In 1996, a tabloid encouraged telephone users in offices to sanitize office phones to prevent getting coworker's phones.[93] One contemporary scholar has theorized telephony as an inherently schizophrenic state of consciousness.[94] Currently, the *Sharper Image* catalog sells a cell phone sanitizer that uses ultraviolet light.[95]

In light of this persistence, I turn in the next chapter to the cultural roots that explain such technopathologies as reasonable and rational.

Notes

1. Kittler, *Gramophone*, 255.
2. Mills, *Deafening*, 124.
3. Kittler, *Gramophone*, 136, 200, 203; 159; 37; 51; 30.
4. Kittler, *Gramophone*, 81.
5. Kittler, *Discourse*, 195
6. Gitelman, *Always*, 3, 10.
7. Hilmes, *Radio Voices*.
8. Morley, *Television*.
9. Mosco, *Digital*.
10. Bottomore, "The Panicking Audience?," Gunning, "An Aesthetic," Sconce, *Haunted Media*.
11. Biocca, "Active," Morley, *Television*, 17.
12. Allor, "Relocating."
13. Morris and Ogan, "Internet."
14. Napoli, "Revisiting."
15. Webster, "Audience."

16. Marwick and boyd, "Tweet."
17. Rettig, "Addiction," ¶1.
18. "User."
19. Anderson, "John Arderne," 1065.
20. Oudshoorn and Pinch, *How Users Matter*.
21. "End-User," ¶1.
22. Ibid. ¶2.
23. Benson, "Field Study."
24. Ibid. 35.
25. Ibid. 41.
26. Corn, *Unfriendly*.
27. Card and Moran, "User Technology."
28. Ibid. 493.
29. Ibid. 494.
30. Ibid. 415.
31. Dillon and Watson, "User Analysis," 16.
32. Ibid. 16.
33. Crary, *Techniques*, *Suspension*; Currell, "Streamling," Sterne, *Audible Past*.
34. Scott, "Postfeminist User."
35. Griffen-Foley, "Tit-Bits."
36. Friedberg, *Virtual*.
37. Jørgensen, "User Interface Issue"
38. Virilio, *Art*, 40.
39. Harvey, *History*, 19, emphases original.
40. Ibid. 2.
41. Scott, "Postfeminist User."
42. Ibid. 3.
43. Nye, *American*.
44. Noble, *Religion*.
45. E.g., Acland, *Residual*, Chun, "Introduction;" Gitelman, *Always*; Gripsrud, "Broadcast"; Gitelman and Pingree, *New Media*.
46. Bermejo, "Audience."
47. Briziarelli, "Dialectics."
48. Gershon, "Un-friend."
49. Hayward, "ATMs."
50. Forrest, "Controversy."
51. "About Web 2.0 Summit."
52. "About O'Reilly, ¶1.
53. McManus, "Web 2.0," Summary, ¶1.
54. McManus, "Web 2.0," Remixing Content, ¶5, emphasis original.
55. Kelly, "We are the Web," headline/subhead.
56. Rawlinson, "Will the 2008."
57. Andrejevic, *Reality*, *iSpy*, Hassan, "Globalization," Jarrett, "Interactivity," O'Brien, Osbaldiston, and Kendall, "ePortfolios," Robins and Webster, *Technoculture*, Scott, "Bubble," Warnick, *Critical Literacy*, *Rhetoric Online*.

58. Thornham and McFarlane, "Claiming."
59. Dean, *Publicity's*, *Democracy*.
60. Rheingold, *Smart Mobs*.
61. Rafael, "Messianic"
62. Foucault, *Abnormal*.
63. Appadurai, *Fear of Small Numbers*.
64. Plotnick, "Touch," 58
65. Ibid. 53.
66. "Wireless Injury to Health," 1353.
67. Mills, "Deaf Jam," "Deafening."
68. Ibid. "Deafening," 122.
69. Ibid. 124.
70. Stern and Gwathmey, *Once Upon*, 61.
71. Ibid. 67.
72. Attali and Stourdze, "The Birth of the Telephone."
73. Stern and Gwathmey, *Once Upon*, 24.
74. Fischer, *America Calling*.
75. J. B., "Telephone Manners."
76. *Dial Comes to Town*, *Now You're Talking*; *How to Use the Dial Phone*.
77. Fischer, *America Calling*, 79, 81.
78. See discussions in Treichler, "Escaping;" McRobbie, *Uses*, *Aftermath*.
79. Hartman, Kraut and Tilly, "Historical", 27.
80. Biggs, "Pleasure Telephone," 44.
81. Marvin, *Old Technologies*, 132.
82. Killen, *Berlin*.
83. Sterne, *Audible Past*, 272.
84. Weiss, *Phantasmic Radio*, 29.
85. Marvin, *Old Technologies*; Tomes, "Epidemic Entertainments."
86. "Echoes and News: Diseases from Telephones."
87. Brox, *Brilliant*, "Electrical Appliances," 58.
88. Hansen, "Early Silent Cinema"; Kirby, *Parallel Tracks*; Sklar, *Movie-made America*.
89. Stern and Gwathmey, *Once Upon*, 60.
90. "Look Out," ¶2–3.
91. Both in Snowden, "Reporting by Phone."
92. Stern and Gwathmey, *Once Upon*, 70.
93. Morgan, "Don't let co-workers"
94. Ronell, *Telephone*.
95. "UV Cell Phone Sanitizer."

References

"About O'Reilly." n.d. *O'Reilly Media*. http://oreilly.com/about. Accessed April 29, 2009.

"About Web 2.0 Summit." *Web2Con.com*, 2007, April 12. http://www.web2con.com/pub/w/62/about.html. Accessed April 29, 2009.

Acland, Charles R. "The Swift View: Tachistoscopes and the Residual Modern." In *Residual Media*, edited by Charles R. Acland, University of Minnesota Press, 2007.

Allor, Martin. "Relocating the Site of the Audience." *Critical Studies in Mass Communication*, vol. 5, 1988, pp. 215–233.

Anderson, William. "John Arderne and His Time." *The Lancet*, vol. 2, 1897, p. 1065.

Andrejevic, Mark. *iSpy: Surveillance and Power in the Interactive Era*, University Press of Kansas, 2007.

Andrejevic, Mark. *Reality TV: The Work of Being Watched*, Rowman and Littlefield Publishers, Inc., 2003.

Appadurai, Arjun. *Fear of Small Numbers: An Essay on the Geography of Anger*, Duke University Press, 2006.

Attali, Jacques, and Yves Stourdze. "The Birth of the Telephone and Economic Crisis: The Slow Death of Monologue in French Society." In *The Social Impact of the Telephone*, edited by Ithiel de Sola Pool, The MIT Press, 1977, pp. 97–111.

Benson, David H. "A Field Study of End User Computing: Findings and Issues." *MIS Quarterly*, 1983, pp. 35–45.

Bermejo, Fernando. "Audience Manufacture in Historical Perspective: From Broadcasting to Google." *New Media & Society*, vol. 11, no. 1–2, 2009, pp. 133–154.

Biggs, Asa. "The Pleasure Telephone: A Chapter in the Prehistory of Media." In *The Social Impact of the Telephone*, edited by Ithiel de Sola Pool, The MIT Press, 1977.

Biocca, Frank A. "Opposing Conceptions of the Audience: The Active and Passive Hemispheres of Mass Communication Theory." *Communication Yearbook*, vol. 11, no. 648, 1988, pp. 51–80.

Bottomore, Stephen. "The Panicking Audience?: Early Cinema and the 'Train Effect'." *Historical Journal of Film, Radio, and Television*, vol. 19, no. 2, 1999, June, pp. 177–216.

Briziarelli, Marco. "The Dialectics of Voluntariat 2.0: Producing Neoliberal Subjectivity through Facebook." *Sociologia del Lavoro*, 2014. http://www.francoangeli.it/Riviste/Scheda_Rivista.aspx?IDarticolo=50607.

Brox, Jane. *Brilliant: The Evolution of Artificial Light*, Houghton Mifflin, 2010.

Card, Stuart K., and Thomas P. Moran. "User Technology: From Pointing to Pondering." In *Readings in Human–Computer Interaction*, edited by Ronald M. Baecker, Jonathan Grudin, William A. S. Buxton and Saul Greenberg, Morgan Kauffman Publishers, 1995.

Chun, Wendy Hui Kyong. "Introduction: Did Somebody Say New Media?" In *New Media, Old Media: A History and Theory Reader*, edited by Wendy Hui Kyong Chun, and Thomas Keenan, Psychology Press, 2006.

Corn, Joseph. *User Unfriendly: Consumer Struggles with Personal Technologies, from Clocks and Sewing Machines to Cars and Computers*, Johns Hopkins University Press, 2011.

Crary, Jonathan. *Suspensions of Perception: Attention, Spectacle, and Modern Culture*, The MIT Press, 2001.

Crary, Jonathan. *Techniques of the Observer: On Vision and Modernity in the Nineteenth Century*, MIT Press, 1990.

Currell, S. "Streamlining the Eye: Speed reading and the Revolution of Words, 1870–1940." In *Residual Media*, edited by Charles R. Acland, University of Minnesota Press, 2007.

Dean, Jodi. *Democracy and Other Neoliberal Fantasies: Communicative Capitalism and Left Politics*, Duke University Press, 2009.

Dean, Jodi. *Publicity's Secret: How Technoculture Capitalizes on Democracy*, Cornell University Press, 2002.

Dial Comes to Town. Dir. Loucks and Norling, AT&T/Bell, Western Electric Recording, late 1940s.

Dillon, Andrew, and Charles Watson. "User Analysis in HCI—the Historical Lessons from Individual Differences Research." *International Journal of Human-Computer Studies*, vol. 45, no. 6, 1996, pp. 619–637.

"Echoes and News: Diseases from Telephones." *Medical News (1882–1905)*, vol. 74, no. 8, 1899, p. 244.

Electrical Appliances 1909. Baltimore, MD: Baltimore Electrical Supply Company.

"End-User." *Investopedia*. http://www.investopedia.com/terms/e/end-user.asp. Accessed Jan 25, 2015.

Fischer, Claude S. *America Calling: A Social History of the Telephone to 1940*, University of California Press, 1994.

Forrest, B. "Controversy about Our 'Web 2.0' Service Mark." *O'Reilly Radar*, 2006, May 25. http://radar.oreilly.com/archives/2006/05/ controversy_about_our_web_20_s.html Accessed April 29, 2009.

Foucault, Michel. *Abnormal: Lectures at the Collège de France, 1974–1975*, Picador, 2004.

Friedberg, Anne. *The Virtual Window: From Alberti to Microsoft*, The MIT Press, 2006.

Gershon, Ilana. "Un-Friend My Heart: Facebook, Promiscuity, and Heartbreak in a Neoliberal Age." *Anthropological Quarterly*, vol. 84, no. 4, 2011, pp. 865–894.

Gitelman, Lisa. *Always Already New: Media, History, and the Data of Culture*, The MIT Press, 2006.

Gitelman, Lisa, and Geoffrey B. Pingree, eds. *New Media, 1740–1915*, The MIT Press, 2003.

Griffen-Foley, Bridget. "From Tit-Bits to Big Brother: A Century of Audience Participation in the Media." *Media, Culture & Society*, vol. 26, no. 4, 2004, pp. 533–548.

Gripsrud, Jostein. "Broadcast Television: The Chances of Its Survival in a Digital Age." In *Television after TV: Essays on a Medium in Transition*, edited by Jan Olsson and Lynn Spigel, Duke University Press, 2004.

Gunning, Tom. "An Aesthetic of Astonishment: Early Film and the (In) credulous Spectator." In *Film Theory: Critical Concepts in Media and Cultural studies*, edited by Philip Simpson, Andrew Utterson, and Karen J. Shepherdson, Vol. 3. Taylor & Francis, 2004.

Hansen, M. "Early Silent Cinema: Whose Public Sphere?" *New German Critique*, no. 29, 1983, pp. 147–184.

Hartman, H. I., R. E. Kraut, and L. A. Tilly. "Historical Patterns of Technological Change." In *Computer Chips and Paper Clips: Technology and Women's Employment, Volume I.*, edited by Heidi I. Hartmann, pp. 24–61, National Academies Press, 1986.

Harvey, David. *A Brief History of Neoliberalism*, Oxford University Press, 2005.

Hassan, Robert. "Globalization: Information Technology and Culture within the Space Economy of Late Capitalism." *Information, Communication & Society*, vol. 2, no. 3, 1999, pp. 300–317.

Hayward, Mark. "ATMs, Teleprompters and Photobooths: A Short History of Neoliberal Optics." *New Formations*, vol. 80, no. 80, 2013, pp. 194–208.

Hilmes, Michelle. *Radio Voices: American Broadcasting, 1922–1952*, University of Minnesota Press, 1997.

How to Use the Dial Phone. Dir. AT&T/PT&T Bell System. 1927, USA.

Jarrett, Kylie. "Interactivity Is Evil! A Critical Investigation of Web 2.0." *First Monday, vol. 13, no.* 3, 2008. http://firstmonday.org/article/view/2140/1947. Accessed July 24, 2010.

J. B., Mrs. "Telephone Manners Are Not What They Used to Be." *National Enquirer*, 1996, January 16, p. 29.

Jørgensen, Anker Helms. "A User Interface Issue in 1956: Preserve the Keystrokes in Usable Form." In *Proc. 7th Danish HCI Symposium*, IT University of Copenhagen, pp. 19–20, 2007.

Kelly, K. "We Are the Web." *Wired*, 2005. http://www.wired.com/wired/archive/13.08/tech_pr.html. Accessed Sept. 12, 2009.

Killen, Andreas. *Berlin Electropolis: Shock, Nerves and German Modernity*, University of California Press, 2006.

Kirby, Lynn. *Parallel Tracks: The Railroad and Silent Cinema*, Duke University Press, 1997.

Kittler, F. A. *Discourse Networks, 1800/1900*, Stanford University Press, 1990.

Kittler, F. A. *Gramophone, Film, Typewriter*, Stanford University Press, 1999.

"Look Out for the Electric Storms to be Very Bad This Summer." *The Manning Times*, 1913, June 25, p. 6. Library of Congress Chronicling America database.

Marvin, Carolyn. *When Old Technologies were New: Thinking about Electric Communication in the Late Nineteenth Century*, Oxford University Press, 1988.

Marwick, Alice E., and danah boyd. "I Tweet Honestly, I Tweet Passionately: Twitter Users, Context Collapse, and the Imagined Audience." *New Media & Society*, vol. 13, no. 1, 2011, pp. 114–133.

McManus, R. 2007, April 2. "Web 2.0 for Designers." *Digital Web Magazine*, April. http://www.digital-web.com/articles/web_2_for_designers/. Accessed May 25, 2009.

McRobbie, Angela. *The Uses of Cultural Studies*, Sage, 2005.

McRobbie, Angela. *The Aftermath of Feminism: Gender, Culture and Social Change*, Sage, 2009.

Mills, Mara. "Deaf Jam From Inscription to Reproduction to Information." *Social Text*, vol. 28, no. 1, 2010, pp. 102, 135–158.

Mills, Mara. "Deafening: Noise and the Engineering of Communication in the Telephone System." *Grey Room*, no. 43, 2011, pp. 118–143.

Morgan, Christopher. "Don't Let Co-workers Give You Their Colds." *National Enquirer*. 1998, January 8, p. 23.

Morley, David. *Television, Audiences and Cultural Studies*, Routledge, 2003.

Morris, Merrill, and Christine Ogan. "The Internet as Mass Medium." *Journal of Computer-Mediated Communication*, vol. 1, no. 4, 1996.

Mosco, Vincent. *The Digital Sublime: Myth, Power, and Cyberspace*, MIT Press, 2004.

Napoli, Philip M. "Revisiting 'Mass Communication' and the 'Work' of the Audience in the New Media Environment." *McGannon Center Working Paper Series*, no. 24, 2008.

Noble, David F. *The Religion of Technology: The Divinity of Man and the Spirit of Invention*. New York: Alfred A. Knopf, 1997.

Nye, David E. *American Technological Sublime*, The MIT Press, 1994.

O'Brien, Peter, Nick Osbaldiston, and Gavin Kendall. "ePortfolios and eGovernment: From Technology to the Entrepreneurial Self." *Educational Philosophy and Theory*, vol. 46, no. 3, 2014, pp. 284–295.

Oudshoorn, Nelly E. J., & Pinch, T. *How Users Matter: The Co-construction of Users and Technologies*. Cambridge, Massachusetts: MIT Press, 2003.

Plotnick, Rachel. "Touch of a Button: Long-Distance Transmission, Communication, and Control at World's Fairs." *Critical Studies in Media Communication*, vol. 30, no. 1, 2013, pp. 52–68.

Rafael, Vincente L. "The Cell Phone and the Crowd: Messianic Politics in the Contemporary Philippines." In *New Media Old Media: A History and Theory Reader*, edited by Wendy Hui Kyong Chun and Thomas Keenan, Routledge, 2006.

Rawlinson, Linnie. "Will the 2008 USA Election Be Won on Facebook?" *CNN.com*, 2008, Oct. 13.

Rettig, Cynthia. "Is Addiction a Useful Concept for Media Use?" *Digital Athena*. 2011, May 26.

Rheingold, Howard. *Smart Mobs: The Next Social Revolution, Transforming Cultures and Communities in the Age of Instant Access*. Cambridge, MA: Perseus, 2002.

Robins, Kevin and Frank Webster. *Times of the Technoculture: From the Information Society to the Virtual Life*, Routledge, 1999.

Ronell, A. *The Telephone Book: Technology, Schizophrenia, Electric Speech*, University of Nebraska Press, 1989.

Sconce, Jeffrey. *Haunted Media: Electronic Presence from Telegraphy to Television*. Durham, NC: Duke University Press, 2000.

Scott, D. Travers. "Bubble 2.0: Organized, Online Critique of 'Web 2.0'." *Rocky Mountain Communication Review*, vol. 6, no. 1, 2009, pp. 32–39.

Scott, D. Travers. "The Postfeminist User: Feminism and Media Theory in Two Interactive Media Properties." *Feminist Media Studies*, vol. 10, no. 4, 2010, pp. 457–475.

Sklar, R. *Movie-made America: A Cultural History of American Movies*, Vintage, 1994.

Snowden, C. "Reporting by Phone." In *Residual Media*, edited by Charles R. Acland, University of Minnesota Press, 2007.

Stern, Ellen, and Emily Gwathmey. *Once Upon a Telephone: An Illustrated Social History*, Harcourt Brace and Company, 1994.

Sterne, Jonathan. *The Audible Past: Cultural Origins of Sound Reproduction*, Duke University Press, 2003.

Thornham, Helen, and Angela McFarlane. "Claiming Content and Constructing Users: User-Generated Content and BBC Blast." In *Youth Cultures in the Age of Global Media*, edited by David Buckingham, Sara Bragg and Mary Jane Kehily, Palgrave Macmillan, 2014, pp. 186–201.

Tomes, N. "Epidemic Entertainments: Disease and Popular Culture in Early-twentieth-century America." *American Literary History 14* No. 4 (2002): 625–652.

Treichler, Paula. "Escaping the Sentence: Diagnosis and Discourse in 'The Yellow Wallpaper'." *Tulsa Studies in Women's Literature*, vol. 3, nos. 1–2 (Spring–Autumn, 1984), pp. 61–77.

"User." *Oxford English Dictionary* online. Accessed Jan 25, 2015.

"UV Cell Phone Sanitizer." *The Sharper Image*. http://www.sharperimage.com/si/view/product/UV-Cell-Phone-Sanitizer/200042. Accessed June 8, 2013.

Virilio, Paul. *Art and Fear*, Continuum, 2000.

Warnick, Barbara. *Critical Literacy in a Digital Era: Technology, Rhetoric, and the Public Interest*, Routledge, 2001.

Warnick, Barbara. *Rhetoric Online: Persuasion and Politics on the World Wide Web*, Peter Lang, 2007.

Webster, James G. "The Audience." *Journal of Broadcasting & Electronic Media*, vol. 42, no. 2, 1998, pp. 190–207.

Weiss, A. S. *Phantasmic Radio*, Duke University Press, 1995.

"Wireless Injury to Health." *Lancet*, no. 1353, 1925.

· 3 ·

NOT SO CRAZY

Electrical Logics of Technopathologies

In 2007, a 37-year-old dentist was jogging through a park in Burnaby, British Columbia, Canada. A thunderstorm was brewing, and lightning struck a nearby tree, throwing him eight feet. His injuries were far worse than would be expected due to his medical condition of "disrupted flashover effect." A healthy flashover effect refers to the skin's natural ability to repel electricity. However, as reported in the *New England Journal of Medicine* and relayed in North American print and broadcast media, the man's bioelectrical state was unbalanced. Instead of repelling electricity, it channeled electricity into his head, resulting in second-degree burns to his chest and leg, additional burns to his head, neck, face, and hip; ruptured eardrums, dislocated bones of the middle ear, a dislocated jaw broken in four places, and hearing damage. The cause of disrupted flashover effect—the factor that made him overly vulnerable and susceptible to the force of lightning—was his wearing the most popular personal digital music player, an iPod. The shape of his earbuds and wires were burned into the sides of his head, neck and chest in a wishbone shape.[1]

Disrupted flashover effect was not technophobic paranoia. Digital music players had been available in the United States since 1998, and the iPod had been available for six years. The story was reported in one of the most prestigious peer-reviewed medical journals. When picked up in popular media, reporters identified a similar case from a year prior: a 19-year old Colorado man had been mowing the lawn while listening to Metallica on his iPod. Lightning struck an adjacent tree. The shape of his headphones was burned into the side of the young man's face, ears, and hip, both eardrums were ruptured, and bones in his right ear were damaged. After two surgeries, he still suffered from mild hearing loss. These two cases were linked to additional ones in which users of portable electronic devices were struck by lightning, or a nearby strike jumped to them. Although such devices do not attract lightning, like a tree or lightning rod, if lightning does strike, the metal in them channels the electricity, causing contact burns and injuries from muscle contractions. More recently, lightning susceptibility has been attributed to the selfie sticks attached to smartphones.[2]

The same phenomenon occurs with jewelry and coins.[3]

As this episode suggests, some technopathologies are painfully real. However, rather than embarking on a quixotic effort to determine which ones are real, or to what degree, in this chapter I examine the cultural roots that underlie all technopathologies. Given that my focus is on the reality of all technopathological discourses—which, from an agential realist perspective, are not "merely"' discourse but entangled, material-discursive practices—here I ground my argument for the significance of technopathologies by showing how technopathologies are far from ridiculous or irrational. This chapter will present an extended historical discussion of the element joining the technological and the biological: electricity. Drawing on primary and secondary sources, I will examine electricity's historic and ongoing roles in health and illness across sites such as spiritualism, electrotherapy, and medical science. There are longstanding cultural precedents for thinking pathologically about technology, which are often forgotten due to more recent processes of separating and categorizing discrete realms of knowledge and practice, formerly connected and even united.

Technopathologies increase after a series of cultural developments. Electricity becomes demystified from the philosophical and spiritual associations of ether, and begins to pervade lives. "Technology" itself has become a thing. Realms of biological and mechanical knowledge become separated enough to allow for coherent cause-and-effect relationships. Disease evolves from evil spirits and divine retribution to imbalanced humors and miasmas, then to discrete, specific entities—germs—that act in cause-and-effect relationships with and associations among people. As Tomes has described, the rise of germ theory instigated new senses of personal and collective responsibility for disease that become intensified in the rise of neoliberal risk society.[4] In contrast to earlier neurasthenia, one does not recover from the deleterious impact of modern society, one preventatively adapts to modern society to take responsibility for not getting sick. At this point, then, various pathologies appear associated with users of electric communication technologies.

In exploring this history, I seek to unpack the thingification of bioelectrical processes so pathological technoculture can be better understood as entangled, productive processes rather than cause-and-effect reactions between things. From this perspective, technopathologies are not irrational reactions of users to technologies, but their mutually entangled, ongoing processes of becoming.

We Have Always Been Electric

The historical work of Carolyn Thomas de la Peña on energy and the human body has shown how Americans' increased energy consumption, production, storage, and distribution in the mid-nineteenth-to-twentieth centuries was entangled with corporeal experiences and understandings. Powering the nation's advance through industrialization, electrification and early radioactivity intersected with bodily power, fitness, and gender in areas such as electrotherapeutics, electrical quack medicines, and pseudoscientific devices. Mechanical and bodily powers were not only parallel and at times conflated, but transferable, increasing human potential. "There was reason to believe," she writes, "in the limitless energetic future of modern men and their machines."[5]

Her book, *The Body Electric: How Strange Machines Built the Modern American*, has been an important contribution to technology historiography and American Studies. By putting human bodies at the center of technology scholarship, she works to engage in a humanistic project that continues the work of resting technological determinism in technology scholarship and popular thought. As such, I see my project as continuing that work. In this chapter, I will engage with a concept key to her work as well: the entanglement of biological and technological energies. As she describes,

> In their attempts to understand the human body, the nature of disease, and the properties of electric force, mid-nineteenth-century American physicians left much room for readers to conflate physical energy with electric power. Even as early as the pre-Civil War era, Americans had begun to extend electrotherapy's potential toward the supernatural.[6]

However, where I diverge from de la Peña is in describing this as conflation. Conflation as a concept is predicated on separation; it is the act of mistakenly or inappropriately considering two separate things as the same. Biological and mechanical energies can only be conflated if their true essences are distinct things. As with Marvin's notion of electrical discourses being shaped to the body, de la Peña's perspective of voluntarily connecting bodies to machines presumes their initial separation. As a queer project, I operate from a position suspicious of true essences and, in drawing on agential realism, am striving to conceive of my phenomena under study as ongoing, entangled processes rather than separate things. Connecting bodies to electricity only makes sense from a modern presumption that they are separate things to begin with. It perpetuates a determinist, utopian/dystopian reaction narrative rather than examining these phenomena as illustrations of processes that were already entangled.

I argue here that these phenomena were always already connected and, indeed, this is why pathological technoculture is neither technophobic nor otherwise irrational. Drawing on primary and secondary historical sources, in this chapter I present a historical overview of well-charted terrain—electrical history—but focused to highlight ongoing entanglements of electricity, bodies, health, spirit, and mind. The combination of technological and pathological seems odd only from a contemporary perspective that separates those realms; this chapter seeks to denaturalize that contemporary perspective.

From Soul to Commodity

What courses through the veins of users? What is the lifeblood of communication technologies? Today, the varied forces that travel through our nervous systems, shoot down from clouds, or surge out of outlets are generally understood as distinct, each possessing unique properties and, more importantly, having been successfully channeled into separate functions. This has not always been the case.

If electricity's ancestors included pneuma—the ethereal, divine force of life and death that created our world—then it was common to physical, spiritual, and psychic realms. Moreover, if, as many scientists and philosophers of the past believed, divinity is expressed through unity, constancy, and commonality, then the forces animating our thoughts, our sexual magnetism, our brains, our bodies, our ectoplasmic spirits, our light bulbs, our death penalties, our medical treatments, and our wireless communications were one and the same, or, at least, very close relatives. Given such ideologies of equivalency and universalism, it was common sense that invisible actions and agents (thoughts, willpower, souls, spirits, demons, emotions) could travel through invisible media (pneuma, ether, lightning, electricity, brain waves, nervous systems, dynamos, radio waves, TV signals, power lines, wi-fi networks, cell phone signals). As ether enlivened the mortal world, electricity animated modern communication technologies, and electromagnetic waves were the ethereal element filling the atmospheric void.

In contrast to Marvin's description of electrical discourses being shaped to and joined with those of the body, I present here the history of electricity as a history of separation, of taking a single fundamental element and refining, dividing, and redefining it into the many distinct things we know today. Including us. Several historians of science and technology consider electrification a fundamental process in the development of modernity and its attendant cultures and subjectivities.[7] The division of our fundamental matter and soul into a utility, a commodity, surely contributes to our understandings of selves and humanity. This is intensified as we increasingly mediate ourselves and connect to others using electric communication technologies. Electricity animates devices often connected with *ether*-net networking technology.[8]

This is not to argue that pneuma, ether, animal spirits, magnetism, souls, lightning, mesmeric force, charisma, telepathy, willpower, static electricity, alternating current, or neural charges were or are all truly one and the same, or that science or culture exerted an effect that transformed one into another. But, these various concepts from religion, science, philosophy, physics, and medicine have been connected. Your soul is not the same thing as or caused by the wall current in an electrical outlet, but being a human who thinks about having a soul is not completely disconnected from channeling energy into mechanical devices. They can be, however, perceived and better understood as general energies electrical, spiritual, neurological, and climatological engaged in ongoing entanglements, the material-discursive practices that gird the phenomena of technopathologies. Moreover, this lineage supports my argument for a significant degree of historical continuity in pathological technoculture. As I will discuss in the subsequent chapter, technopathologies are not solely associated with new or emerging technologies. They persist, and engaging with the cultural history of electricity will clarify some reasons why.

Creating Things, and Categories for Them

Over the nineteenth and into the early twentieth centuries, processes of division and categorization in Western culture culminated such that natural forces, formerly considered entwined, integrated, or equated, became separated as distinct fields of knowledge and natural phenomena. Specifically, the electricity running through our brains and muscles was reconceived as distinct from that shooting from the sky or zipping through wires. What was once the breath of the gods became a commodity; the essence of our souls became the essence of modern communication technologies, among many other things.

Today, we have arrived at an understanding that distinguishes *bio*-electricity as a separate thing from that which powers our computing devices. Living tissues, not meteorological phenomena or manipulation of natural elements, create bioelectricity. Bioelectricity generates the same potentials as generators or batteries, but bioelectricity is a flow of electrically charged atoms or molecules (called ions), whereas household electrical current is a flow of charged subatomic particles, electrons. Although significant, this technical distinction is perhaps not widely understood by the general public. It is also only somewhat recently that this force has been separated into specific, distinct types. Historically, it first took centuries to understand that the force that shot down from the sky in lightning was the same force that allowed the mind to command muscles to move, and then was captured, controlled, and used for tasks such as illuminating a light bulb.

It has been argued that the division of the world into categories is fundamental to human psychology and intelligence, and especially foundational in Western culture and thought. Categories separate; they create divisions. In this sense, they reinforce an object-centered view of the world: X and Y are different *things*. Moreover, they typically imply exclusivity—if X, then not Y. Clearly, categories can be tools of inequality and conflict, but they can also be productive, for they allow, not only acts of identification and grouping, but also acts of making visible the relations between distinct objects. If a person is in the category of "monster," then they are one and the same, so what can one do? That person *is* a monster, they are one and inseparable. But, separation allows for relationships. If a virus or gene causes a person to appear monstrous, we can intervene in that relationship to effect a cure.

As electricity was moving from the heavens and our bodies to storage batteries and mechanical devices, other processes of separation and distinction were also occurring in medical knowledge. Disease was changing from a model of humors and miasmas to that of germs: ephemeral, innocuous fluids and gases became specific, harmful things. We human beings, ourselves, were also splitting, as formerly unitary models of inherent essence and character gave way to psychiatric models that split us into minds and bodies, and illnesses, therefore, into categories of mental and physical sickness. Ed Cohen describes this as the *modern body*.[9] In his genealogy of the concept of biological immunity, Cohen argues that it forms part of a biopolitical discourse that helps constitute a modern body, one that is defensive and separate, distinct from its surrounding lifeworld and other modern bodies. It is one of separation and thingification, and it evokes the separation of human from inscription that Kittler describes in the mediality of modern discourse networks. However, as I suggested previously, such epochal historiography focuses on rupture and change, whereas I am trying to employ an agential realist perspective that sees ongoing, entangled processes. The question then becomes, not what changes, but how are different phenomena entangled in their ongoing, iterative processes of becoming?

These processes of separation and distinction created new things, new objects. New things enabled new actions. In the Marxist concept of objectification or reification, abstract human attributes, relations, or actions, such as will, labor, or subjectivity, are invested in man-made things, thereby becoming objects. The value of a worker's labor is turned into the object of a gold coin. One can now act upon this object in ways one couldn't have acted upon the abstract value associated with the original, actual labor: it can now be traded, spent, or saved.

Typically, this process, sometimes also referred to as "thingification," is described critically, as a component in processes of capitalistic alienation and exploitation. For now, however, I am using it more functionally, to call attention to how processes of separation and distinction create objects, or new things. Separateness and distinction enable certain possibilities. They create the possibility for relationships. Things can relate. A unitary being or phenomenon does not relate, but two or more can act upon each other, one can cause certain effects upon the other. As part of my project of examining pathological technoculture in a manner that does not reify object-oriented ontologies and linear, cause-and-effect relationships, I must first describe how what are today thought of things were not always so.

It All Starts with Ether

Ancient philosopher-priest-scientists—roles which were once overlapping if not united, only becoming separate and distinct to a modern perspective[10]—divided the world into five elements: earth, water, fire, air, and ether. What was ether? Historian Joe Milutis refers to it as "the nothing that makes up everything," the idea being that the ancients could not abide by the concept of nothing in the way we can think of, for example, the emptiness of outer space.[11] Instead, the planets floated through ether. What, after all, kept them afloat? Like boats on water, they must float on something.

To the ancients, our planet's atmosphere was filled with air, which mortals breathed. But, further up in the realm where the gods lived, they breathed something else—ether, the element that made up their space. Moreover, given that ether was what the gods inhaled, it was also what they exhaled. And it is exactly through their exhalations that the gods had created the mortal world: pneuma, or the breath of life, is divine exhalation of ether. This concept continued in Christianity: "And the LORD God formed man [of] the dust of the ground, and breathed into his nostrils the breath of life; and man became a living soul."[12] Divine respiration inspires life in inanimate matter, and, when this force leaves us, we expire. The gods' exhaled ether turned inanimate rock into our Earth and its lifeforms, and also filled up the space between the earth, sun, and other planets.

Ether is a technology of life- and world-making. It "makes up everything," and technology—particularly divine technology—threatens mortals. Invisible waves of power doom the transportation technologies of Daedalus and Icarus; Prometheus' theft of divine fire brings retribution down on humanity (notably, in some versions, in the gendered form of Pandora); Frankenstein learns his lesson, as do numerous victims of irradiated dinosaurs, insects, and other animals of 1950s atomic horror films.[13] Therefore, it is entirely logical for electric communication technologies to be a site of health concerns. As one member of my focus groups described:

> Computers are giving off electromagnetic energy. Our body's energy field is electromagnetic energy. So somehow or another there must be a meshing of these two energy fields, and if that's the case, does our energy that we give off affect what's going on in the computer? And I've would say that, without a doubt, that if someone sits in front of the computer, and they're in a good frame of mind, the computer is going to run much different than if they're in a very chaotic, scattered place. That computer is going to be bouncing all over.

Note here that technopathological discourse is about the user ("if someone sits in front of a computer and they're in a good frame of mind"), and also the general ease with which even a twenty-first-century user conflates biological and mechanical electricities. Vitalism, physics, electrotherapy, psychoanalysis, spiritualism, hypnosis/mesmerism, vital photography, mind control, and the supernatural are the interconnected and multifaceted roots of what we take for granted today as mere electrical current.

Spirit and the Supernatural

In 1919, the West Coast literary magazine *Overland Monthly* ran two stories of supernatural connections, illustrating a common narrative of spirit communication through electric technologies.[14] In one, the spirit of an unhappy, murdered Native American Chief communicates through telegraph, inciting fear and trembling in operators, and later directly haunts one operator, provoking him to break his leg. In another, merely reading a story about spirit communication through the telegraph drives a telegraph operator into a state of near nervous collapse and anxiety.[15]

Technological spiritualism was perfectly sensible. Spirits could communicate with the living because electrical energy traversed both the realms of the living and the dead. As Jeffrey Sconce and others have described, it was understandable that electrical telegraphs and radios could pick up the voices of ghosts. Electromagnetic waves invisibly connected radio transmitters and receivers across space, so why couldn't they connect to the electrical brainwaves of human minds, in this world or beyond? We still see today similar stories about how photography can capture images of ectoplasm and aura, tape recorders can capture spirit voices, televisions can entice little girls into the spirit-world, videotapes can carry a ghost's viral curse, and even a fax machine can transmit clothes of the dead.[16]

This chart from a book by George Beard, the "inventor" of neurasthenia, presents a variety of scientific, supernatural, and metaphysical fields from a unified perspective (Fig. 3.1).

Although the inner ring represents true science and the outer "delusions," it is significant that they are still conceived of as concentric, adjacent, and connected. They are not separate, distinct fields but different parts of a continuum running from fact to superstition, suggesting indeterminate, liminal areas in-between. Indeed, Foucault argues that the importance of the electrical nervous system in eighteenth century medicine (and arguably, later)

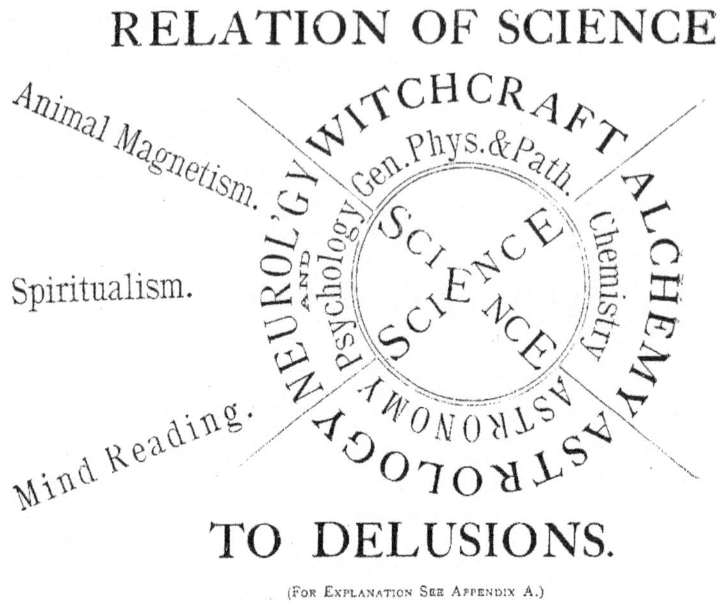

Figure 3.1. "Relation of Science to Delusions." Discourses and disciplines today considered very distinct shown in a more unified, connected perspective.
Source: George Miller Beard. *The Psychology of the Salem Witchcraft Excitement of 1692 and Its Practical Application to Our Own Time*. New York: G. P. Putnam's Sons, 1882. Huntington Library and Archives.

was that it recodified the domain of "concupiscence … the sinful soul of the flesh," which had formerly belonged to the art of Christian penance.[17] In other words, the medicalization of something from a spiritual domain represented a transfer of power in administering that domain. However, I argue that the act of transference does not go unnoticed, and the domains remain linked in many ways in popular culture and habits. Thus, the electricity in communication technologies resonates with supernatural ectoplasm and spiritual ether.

In past decades, tabloids have reported cases of technological spiritual crossovers, such as "Ida Lupino Tells of Phone Call From a Dead Friend," "Randy Quaid: My Dead Dad Caused a Power Outage," "We Were Terrified by Ghost Who Spoke Through TV Set," and "Hellish Phone Call."[18] In many fantastic films, electric communication technologies connect users to dangerous, malevolent spirits. These include social media in *Unfriended*, a fatal website in *Horror Vision*, computers in *Evilspeak*, telephones and telephone numbers in *The Booth*, *976-EVIL*, *976-Evil II: The Astral Factor*, *End Call*, *House on Haunted Hill*, and *A Nightmare on Elm Street*; shortwave radio in *Shortwave*, silent film in *Return to Babylon*, theatrical movie film in *Reel Horror*, home super-8 film in *Sinister*, television in *Nightmare on Elm Street III: The Dream Warriors*, *Dead Waves*, and *Shutter*; closed circuit television in *This House Possessed*, videos in *Playback*, LPs in *Trick or Treat* and *The Lords of Salem*, and a portable cassette stereo in *Amityville: The Possession*. Amateur ham-radio broadcasts are the tool of a young female ghost luring victims to a *Ghosthouse*, the cryptic message turning out to be a recording of their future deaths, prefiguring similar voicemails in the more *One Missed Call* films.[19]

One woman in my focus groups described how technological connection to supernatural forces, even if not malevolent, could still cause mental distress:

> In the town I'm from there's actually a house that, when I was a kid, I used to babysit. … There is a book on it, like, for being a haunted house, and, like, a couple of homes that I've heard of and read documentation on, I know people who lived in and have been written up as haunted houses, like officially, whoever officially takes that. And a lot of stories … involve, like, hearing the TV go on or off in the other room when nobody's in that room, or a light is on in one room while everyone's in the kitchen and like the light just pop, go out. In terms of it being negative, it's just that it puts fear in people. That's like a negative thing. And I guess if it's like supposedly an angry poltergeist, they'll like do more. They'll, like, blow out fuses as opposed to just having some static, because some spirit is entering.

In my focus groups, the film *Poltergeist* was frequently remembered as an example of this. One woman said, "The girl was seeing the ghost coming from the TV, and I guess it was the TV that was—I don't know—possessing her mind or something like that." This is actually incorrect, a mis-remembering that attests to the salience of the general technopathological pattern. In *Poltergeist*, the daughter only initially *communicates* with spirits through staring into television static; she is *abducted* through a dimensional portal growing out of her bedroom closet and terminating in the living room. Moreover, she is never actually possessed. Although kidnapped, she does not take on the consciousness or appearance of another entity. Yet, the film's advertising and final image—the family exiling a TV to outside their motel room—ascribe significance to the TV unsupported by the actual narrative, yet highly resonant with cultural roots of electrical-supernatural connections.

Biology and Physics

The idea of a mostly invisible substance that suffuses the world with powers of life and death has had great persistence, most recognizable today as electricity.

As lightning, it was what Zeus threw down from his ethereal realm, connecting gods and mortals. Enlightened scientist-philosophers, arguably the replacements for the priests of earlier centuries, studied and harnessed lightning from the heavens. Volta demonstrated the reciprocal influences and combined interconnections of mechanical, spiritual, biological, and scientific discourses that today's disciplines and epistemologies have separated. Benjamin Franklin considered electricity universally present in all matter: "I conceive then, that this globe of earth and water, with its plants, animals, and buildings, have, diffus'd throughout their substance, a quantity of electric fluid, just as much as they can contain."[20] In the nineteenth century, universal theories of equivalence to large degrees did not separate mechanical and biological energies. The forces that caused muscle movements and metrological lighting were one and the same. After all, electricity was the common element that connected disparate things, just as ether was a surrounding, supportive, and connecting element. The forces that moved the heavens also moved our muscles, and animated our minds and hearts. Movement, thought, and emotion were all electrical, so were souls, spirits, personality, willpower, and character. Your willpower that coursed down your spinal cord to make a muscle contract was the same force charging brain cells to conceive a thought and inspire mental creativity. This same force that you willed to think a thought was also what was used to exude sexual attraction toward someone, to generate and radiate the charisma of personality.

Today, these many areas have been separated, and the processes of electrical development relate to one of the most famous and influential separations—that between mind and body. It introduced a simple question that would produce many distinct fields of inquiry and knowledge. That question was motion. Specifically, animal motion.[21] If mind and body were separate, then how did one affect the other? How did a thought in one's mind, such as "bend my knee," cause the motion of muscles contracting and a knee bending?

The question of animal motion was a driving force in the development of physics, mechanics, and ultimately, modern electric communications technologies. Ancient and modern thinkers ruminated, meditated, and experimented to divine the workings of this process, variously theorizing intermediary substances that connected mind and matter, or fundamental substances that united them: ether, pneuma, animal spirits, mesmeric force, electricity, willpower, charisma, and vital forces all entered into these discussions in similar and disparate ways far too complex, and employing vastly different terminologies and conceptual frameworks, to review here.[22]

What is important is understanding that the long histories of biology, physics, and even the supernatural were parts of the same universal system, not distinct areas of life and knowledge. For Renaissance scientists, matter had qualities that were both observable ("manifest") and imperceptible ("occult"), although these were debated by Christian thinkers that insisted on observation. However, in the seventeenth century, occult qualities become an accepted part of natural science, with an increasing belief that science could, if not see them, ascertain their workings.[23] For example, the "occult superelemental qualities" of Lisbon physician Duarte Madeira Arrais involved mysterious qualities of elements that could awaken similar qualities, previously unseen, in other substances, such as a lodestone "magnetizing" another object, awakening its inherent force of attraction or repulsion.[24]

Science and spirit—rational observation and faith in the divine—were seen as united pursuits, not contradictory. As one medical historian describes, "The occult qualities or principles of matter, so important for understanding the true nature of God's creation, could only be evinced, it was claimed, by experimental procedures. Thus, the professed belief in occult qualities came to be amalgamated with or embedded into other arguments in defense of the experimental method."[25]

In the eighteenth and early nineteenth centuries, research into electricity integrated biology, physiology, philosophy, medicine, physics, and other areas. Famously, doctor, philosopher, and physicist Luigi Galvani demonstrated that electricity would cause the muscles of a dead frog to twitch, something with which Alessandro Volta, the inventor of the battery, experimented as well. In 1835, an American doctor in Virginia performed similar experiments on the body of an executed African-American slave, which was described at length in an 1859 medical book proposing *Variations of Electric Tension as the Remote Cause of Epidemic and Other Disease*.[26] In a series of experiments, a Dr. Dunbar hooks up a battery to the heart and nerves in the neck, muscles of the neck, chest, stomach, throat, and abdomen, and nerves attached to the brain, diaphragm, and face. Achieving various twitches, contractions, and expressions, Dunbar laments that his results were not as strong as those produced and described by another doctor: "None of those wonderful plays of the features were well marked."[27] Dunbar describes electrifying the tongue as one of the most interesting experiments, in which "a very striking effect was produced … a rapid vibratory motion of the tongue, compared by several gentlemen to that of a serpent's tongue when alarmed or enraged."[28] Other experiments induced teeth to strike together, an action similar to muttering or reading to oneself. A friend's notes read: "'The lips acquired a motion precisely like those of a person reading to himself; and I do not know if this was not the most natural and pleasing experiment,'" to which Dunbar adds, "The expression of countenance of the criminal, in this stage of the experiment, caused a stare of amazement in many of the spectators."[29]

Subsequent experiments created motion in feet, legs, and arms, "It repeatedly made efforts resembling those of life to jerk its hand away from his grasp, and when permitted, struck against the chest with great violence, settling in the attitude of a pugilist when prepared to defend from the attack of an adversary."[30] However, when Dunbar relates that his results were not as extreme as he had seen related in the work of a contemporary, the author of the book explains this as partially due to the contemporary experimenting on "a strong man, with well-developed organs," whereas, with Dunbar, "The case of the negro represents the condition of one that is feeble and ill supplied with vital electricity."[31]

Such a gruesome episode indicates not only the overlapping realms of different scientific discourses, but also their involvement in exploiting and perpetuating racial power hierarchies, as will appear several times in this book. Biology and physical science exploit the body of the executed slave as raw material for experiments, then use the pathologization of racial subjects to justify their exploitation.

As late as 1871, in a book on medical uses of electricity, two prominent physicians discussed medicine, natural physics, and electrical science as separate realms—and yet, at the same time, exhibited a more ethereal logic of association, correlation, and unifying parallelism:

> There would appear to be cogent reasons for the theory that the law of correlation and conservation of forces is as true of disease as of death. … General observation[s], taken together, would seem to give weight to the theory that *the secret forces of diseases are as truly correlated to each other as heat or gravitation, as magnetism or electricity*.[32]

Health and Therapy

Electricity, in its broadest sense, explained causal relations between separate things because it travelled from one to another and connected them. It moved from heavens to earth, from static-charged fingertip to another person, from wire to wire, from transmitter to receiver. It could be stored in batteries and displayed in light bulbs. Electrical shocks could be harnessed to kill a criminal and cure baldness. Therefore, certain associations that were perfectly logical then seem ludicrous today. For example, in the 1859 book, *On the Influence of Variations of Electric Tension as the Remote Cause of Epidemic and Other Disease*, a prominent London surgeon and physician argued "that electric power and nervous power are identical" and diseases were due to "electrical tension," a state of balanced charge in living beings.[33] Imbalanced electrical tension caused malaria, plague, shipboard pestilence, intermittent fever, and other diseases:

> The yellow fever of the African coast, the septic pestilence of Syria and Egypt, and the cholera of all the rest of the world, have in common the same destruction of vital resistance, and reduction of the power of cohesion of the tissues.
>
> Though local causes may be found to account for the distinctive character of each pestilential disease, yet their general manifestations point to one great operating cause that must produce them all; and what other element in nature, more than electricity, could possess such an agency, could exert such a power?[34]

Ten years later, in *A New Theory of Disease; Based upon the Principle that Man is a Compound Electrical Magnet*, W. R. Wells wrote that, "We contend that there is but one grand cause for all disease, and this is, a loss of balance of the two forces of electricity in the part or parts diseased. … There are no substances in nature that can fully resist the decomposing power of a current of electricity."[35] In later decades, Thomas Edison championed electricity as a mode of execution in his support of the electric chair as corporal punishment (using, naturally, not Edison's own direct current, but the alternating current of his competitor, George Westinghouse).

What could sicken could surely cure. The late nineteenth and early twentieth century saw a wave of electrotherapeutics, as it was thought application of electrical current to the body could cure a large number of ills by regulating the body's internal current (Fig. 3.2). Doctors and entrepreneurs claimed to improve and maintain life through a multitude of medicinal electric devices, such as the violet wand. Edison this time, in 1910, produced the first cinematic version of the story of Dr. Frankenstein—perhaps the quintessential modern, Western myth of technology and gender—who, succumbing to masculine hubris and desire for reproduction without women, "infuse[s] a spark of being" into dead flesh.[36] In most subsequent filmic representations of "the modern Prometheus," he harnesses lighting to electrically create life. Even the word "electricity" had medicinal connotations and was applied to tonic medicine. A newspaper in Winsboro, South Carolina reported in 1906 on

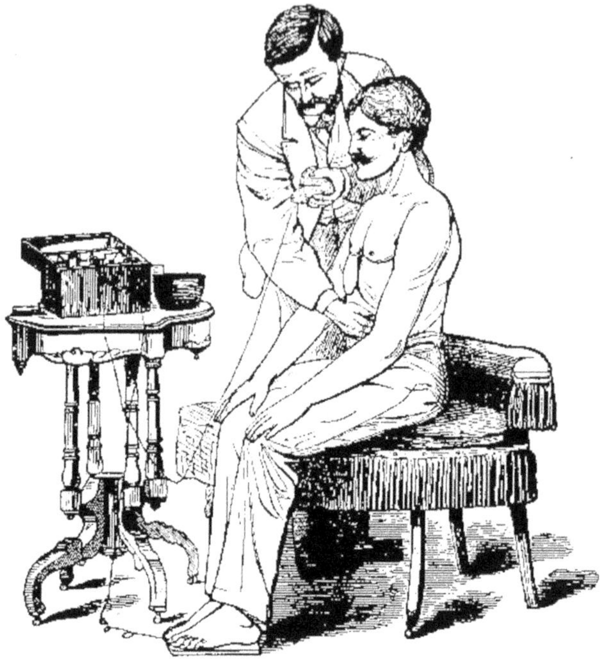

Figure 3.2. "General Faradization—Application to the Lower Extremities." Sponge used to apply electrical current to patient, an example of electrotherapy.
Source: Rockwell, A. D. and George Miller Beard. *A Practical Treatise on the Medical and Surgical Uses of Electricity, Including Localized and General Electrization*. New York: William Wood and Co., 1871, p. 210. Huntington Library Collection.

"Electric Bitters" being successfully used to cure "female troubles" instead of the potentially fatal surgery a doctor had proposed.

Electrotherapeutics extended to mental health. George Miller Beard wrote an extended parallel between the psychology of the 1692 Salem witch trials and the trial of then-contemporary presidential assassin Charles Guiteau: "Both trials were the result of ignorance of the nervous system in disease—that of Guiteau, ignorance of insanity; the Salem trials, ignorance of insanity, hysteria, trance, and allied states, more complex phenomena than those of insanity alone, but all referable to the nervous system."[37] That Beard, a neurologist, expounded on psychology in relation to spiritualism, and rooted all in the biological nervous system, is an example of how less discrete such disciplines and discursive arenas once were. Beard notes, "My studies of the nervous system have many times led me into circles quite as superstitious and grotesque as the circle that constituted the attackers in the witchcraft trials"—something today no psychologist or neurologist would likely say.[38]

Early psychiatry thought the charismatic personality of the doctor literally interacted with and repaired that of the patient in a physical, biological sense, a concept that evolved into the practice of mesmerism, an explicitly electro-spiritual form of intersubjectivity and psychic control (Fig. 3.3).[39]

When this practice became discredited, it was rebranded as hypnotism, a sanitized tool for therapy and anesthesia. Yet bio-electrical connections persevered: nineteenth-century neurologists considered excessive sensitivity to hypnotism a symptom of female hysteria, often performing public displays. In his examination of the U.S. conflation of technology and ethereal spirituality, Jeffrey Sconce suggests that there was an "electropsychopathology" at root in the central nervous system dysfunctions attributed to the bio-electrical-spiritual circuits of nineteenth century "spiritual telegraphy," and these persist in today's conceptions of Internet addictions and mental illness.[40]

Indeed, electrotherapy remains with us today. Although easy to dismiss as archaic naiveté, medical research in the 1980s investigated electricity's role in limb regeneration among amphibians,[41] and the more recent fad for electromagnetic health bracelets indicates how such beliefs persist in popular culture, even after science and medicine have advanced past them. Physical therapists use electrical manipulation to manipulate muscles. Doctors' use of electro-convulsive therapy (ECT) for mental illnesses has gone from celebrated to condemned to respectable in contemporary medicine. The lead character on the *Homeland* television series receives ECT for her bipolar disorder. People from healthcare professionals to flight attendants employ defibrillator paddles to electrically shock the dead back to life, but this is not a case of strict medical need: Tercier argues that this component of CPR has become, in some cases, more cultural ritual than medical necessity, given its uneven success rate.[42]

Processes of Dis-connection: Thingifications of Disease and Technology

The definitions of disease vary, as do understandings of their causes and also the very fields that are considered acceptable sources of knowledge for generating those definitions and understandings. Just as there are many kinds of disease states, there are many ways to become diseased. Some are more popular

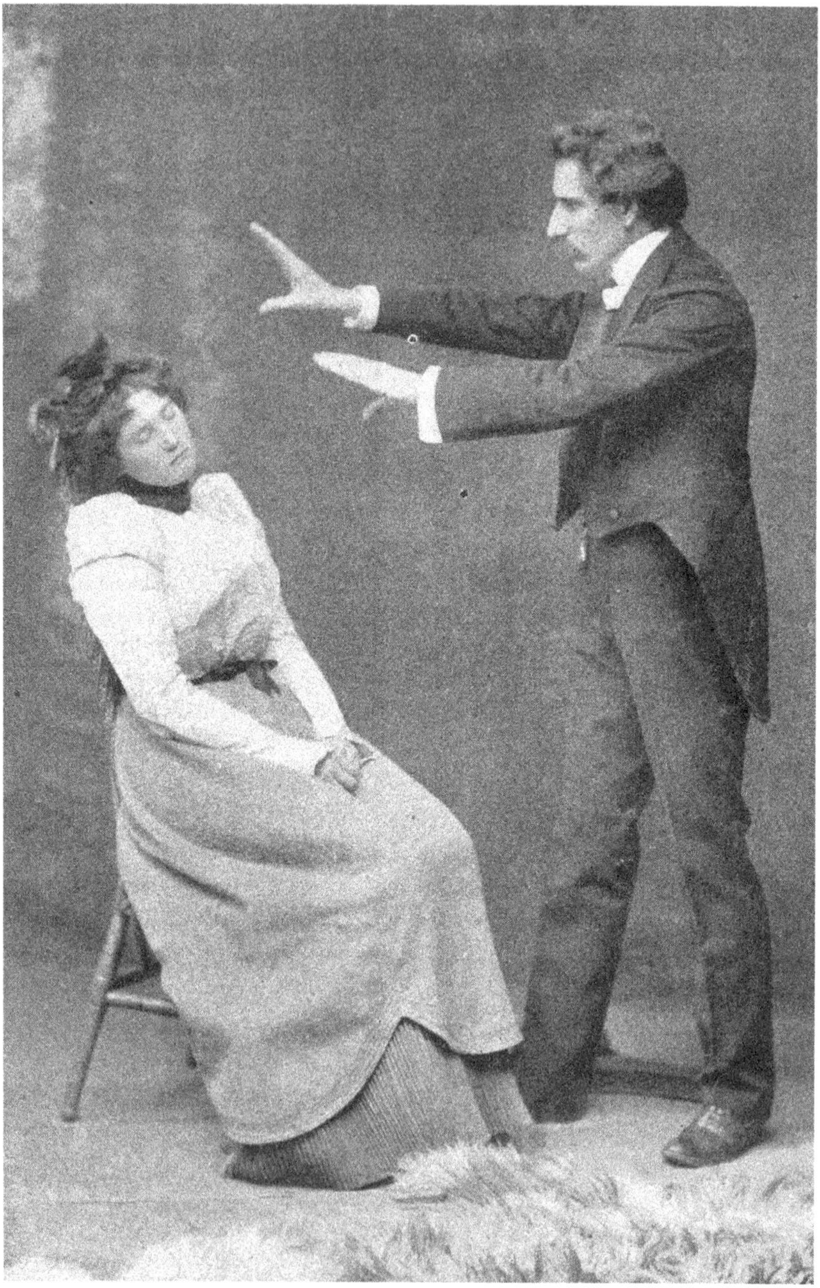

Figure 3.3. "Sleep induced by mesmeric trance." Mesmerism (hypnosis) involved the exchange of energies between parties and was used as a medical anesthesia.

Source: From L. W. de Laurence, *Hypnotism, Magnetism, Mesmerism Suggestive Therapeutics and Magnetic Healing.* The de Laurence Company, Chicago, 1910. Plate between pp. 128–129. Huntington Library Collection.

and legitimated today than others. To fully appreciate technopathologies, one must look beyond narrow, modern conceptions of disease, and beyond comfortingly tidy divisions of phenomena and knowledge. It is important to realize the changes in science and medicine do not always absolutely overthrow their predecessors. Habits, patterns, turns of phrase, and modes of behavior linger.

In the history of disease, logics of association have a much longer presence than those of direct cause. Often more was understood about sickening people and places than essential causes. Medicine has many more centuries of nosology—the identification and categorization of diseases—than etiology—the understanding of what causes mental or physical distress. The mode of communicating syphilis, for example, was known before the disease even had a name, and for four centuries before its etiology was understood.[43] This deep-rooted cultural logic supported both literal and social pathologies. We may not have known what causes syphilis—or AIDS—but we knew what types of people can give them to you, and they are often stigmatized as bad people.

Germ theory—the concept of disease as infection by a bacteria or virus—is a relatively recent perspective.[44] For much longer than they have known what causes them, physicians and scientists have understood diseases as phenomena that can move from one person, place, or thing to another—but it was unclear exactly how this took place. Ancient medical theories included evil spirits, curses, or divine wrath and punishment for evil behavior. Others saw outward physical dysfunctions and disorders as mimetic correspondences with inner immorality or corruption. Often these were combined such that an external factor linked with an internal weakness or predisposition in the sufferer. In classical Greek and Rome, the concept of person-to-person contagion was theorized, evidenced again later by quarantines of lepers in the Middle Ages. Continuing into the eighteenth century, disease states were understood within the concept of humors, the four bodily fluids of blood, phlegm, yellow bile, and black bile, which could be thrown into imbalance through changes in diet and/or activity. These imbalances not only caused diseases but four corresponding temperaments. Note that how, within this model, not only are disease and personality correlated, but disease is also the fault of the sufferer. Improper diet or activities throw their humors out of balance, not invading, infectious agents. This individualizing, pathologizing view of a disease sufferer is one that persists in pathological technoculture, with sick users having suspect characters and bearing responsibility for their own suffering.

Another ancient theory of disease, one common in many parts of the world and that continued into the nineteenth century, was that of miasma.

Unlike humors, miasma theory did involve a foreign agent causing disease: bad air or vapors emanating from decomposing organic matter. This poisonous gas or dirty fog could come from stagnant natural water but also unhygienic urban conditions. Foul-smelling and powerful, it was not merely pollution carried on air but a force that transformed air into something toxic, thought by some to have almost magical properties. Diseases were not spread from person to person, but were carried along in miasmas that sufferers inhaled. In cases such as eighteenth-century cholera epidemic or the famous case of Mary Mallon ("Typhoid Mary") in the early 1900s, this perspective often combined material dirt, filth, and squalor with moral laxity, degeneracy, sloth, and other character failings, often attributed to particular social groups.

As an invisible or evanescent substance that floats through the air, magically charged and affecting life and death, miasma bears direct similarities to ether, the elemental substance entangled across histories ranging from electricity and animal motion to neurology and the supernatural. We see then, in centuries of etiological theory, a substance that behaves in a very similar way to substances that also charged our souls, skies, and technologies. My point is not that any of these things were or are the same, but that similar patterns of thought and behavior appear across histories of disease and technology, which inform an understanding of pathological technoculture.

The current etiological model of infectious disease, germ theory, was not scientifically verified until the later decades of the 1800s, and not without much controversy. It was not until the twentieth century that it became commonly accepted knowledge to believe that invisible living entities were passed from one person to another, causing sickness.

The history of causes of mental illness are separate, yet in many ways parallel. Madness moves from being possession by an evil spirit, a curse, or an inherent personality state, to something else. Like humors, it can be an imbalance caused by an individual's actions: sexual hedonism or repression, exposure to inflammatory ideas, or associating with the wrong types of people. Like miasma, it can also be something invisibly spread across populations, as in cases of mass hysteria. Like germ theory, it can be an outside agent that injures an innocent psyche: childhood trauma, abuse, or the universal traumas theorized in Freudian psychosexual development.

Historian Linda Simon notes that the advents of electrification and the feverishly debated germ-theories of infectious disease were coterminous and mutually informed:

> The theory that illness could be caused by invasive microscopic agents supported the idea that electricity could also be an invasive power. "The 'seeds of disease,' to adopt a popular term (whether we accept the *germ theory* or not)," commented *Scientific American* [in 1884], "are floating about us in myriads without number, and are inhaled by us with every breath, and yet the diseases are manifested only here and there, wherever the 'seed' finds a susceptible point for its growth. In the same manner, though the electrical influence may come alike upon all, yet is its effect made manifest to us in certain cases with great power, while in others we fain to detect it." Physicians noticed that some people were more sensitive to electrical charges than others, reporting tingling of the skin, for example, hours before a thundershower. In these people, electricity could incite nervous symptoms and result in neurasthenia.

Note how, even at this early date, we see user pathologization, a blaming of the user, not the technology.[45] Some people are more sensitive and susceptible to germs and electricity. While not strongly stating it is their own fault, their constitutional weaknesses are individualized, suggesting social hierarchies and divisions.

Thingification also occurs with technology. "Technology" is notoriously hard to define, but as Rosalind Williams remind us, its roots are in *techné*, a common root with *technique*: a way of doing something. She argues that the understanding of "technology" as a thing, rather than a way of doing something, is a modern, twentieth century phenomenon. As she and others have noted, what was formerly a *way* of doing something—technologies of raising crops, for example—became something in and of itself. The concept of "technology" coalesced into a distinct thing, a discrete object of study and knowledge. Previously, saying that you "studied technology" would have sounded as incomplete as saying you "studied processes," each eliciting the question, ". … of what?" Moreover, in recent decades, "technology" itself has evolved or, more accurately, reduced in common usage to refer primarily to technologies of communication. Williams notes this as well, and it is particularly evident when looking through older books on technology. For example, the 1979 edited collection *Technology and Change* includes an annotated appendix of suggested further reading, listing over 150 major twentieth-century works of technology history and study. Of these, strikingly few are explicitly about communication technologies. Instead, areas of interest include engineering, energy, transportation and space exploration, industrialization, pollution, manufacturing, military technologies, scientific research, invention, and economics. Although three address a topic that would later become central to communications—computing—only one is directly related to communication technologies, and, even then, it is not the focus of the listed book, a biography of Thomas Edison, and is described as "an outstanding biography which emphasizes Edison's economic motivations."

Today's understanding of "technology" is relatively recent. However, to better understand contemporary ways in which people describe and react to technology, it is necessary to attempt to wrench free from the myopia of presentism and engage with historic precedents, and those of intersecting areas of knowledge, such as the discourses of health and medicine that this book explores. In other words, the object of "technology," as currently understood, has historic entanglements from which, I argue, it is not completely severed. As with "disease," while we may understand something according to a contemporary definition, it is still rooted in earlier understandings, and there are well-worn paths and patterns with which we engage and understand it. Medical science in the West may no longer consider the nineteenth century affliction of neurasthenia a real disease, but it lives on when we colloquially speak of a nervous breakdown or frayed nerves. Moreover, we may understand disease as, for example, infection by a micro-organism, but we may act sometimes as if it were possession by an evil spirit. We may understand "technology" as a tablet computer, but may act as if it was connected to our bodies and souls.

Appreciating Connections

Technopathologies have effects. Health panics over X-ray-emissions from color TVs and other "electronic products" caused the U.S. Congress to enact Public Law 90–602 in 1968, regulating their output, which remained in effect for four years.[46] Today's concerns over cell phone frequencies continue to fund a mini-economy of research, publication, and consumer protection devices. Horror narratives such as the second episode of BBC television program *Dead Set* suggest wi-fi networks brought about zombie hordes: As one character says, "God knows what that does to our molecules."

In his study of cell-phone health fears, Anthony Burgess argues that late-nineteenth-century fears over health risks from electrical lighting, health panics around X-rays leaking from early color televisions, microwave oven leaks, 1970s fears over power lines and childhood leukemia, conspiracy theories over microwave weapons developed in Cold War secrecy, 1980s labor activism around radiation from video display terminals, and cell phone health concerns all "form part of a general twentieth-century fear generated by the technological harnessing of different parts of the electromagnetic spectrum, even simply electricity itself."[47]

Electricity, and its cousins ether and radiation, are mostly invisible forces, just as germs were for centuries. They are infectious agents of unboundedness, a key element of Mary Douglas' analysis of social fears of pollution. The processes of separation that have made technopathologies conceivable can be appreciated when one looks backwards at how various fields of knowledge and phenomena were once less separate. Just as the different definitions and causes of diseases may change, yet never wholly disappear, more recent processes of separation and distinction are never immediate and absolute. In pathological technoculture, realms we today consider distinct—the biological, mechanical, and supernatural—overlap and interact. This seems less irrationally technophobic and more logical when one takes into account the historic intermingling of these areas, echoes of which persist.

Notes

1. Associated Press, "ER Docs," Bellett, "Shocking," Fitz, "Shocking"; Heffernan, "Thunderstorms," Ryan, "Warning."
2. Matyszczyk, "Selfie."
3. "Could an iPod," Associated Press, "ER Docs," Bellett, "Shocking."
4. Tomes, "Epidemic Entertainments."
5. De la Peña, *Body Electric*, 136.
6. Ibid. 92.
7. Asendorf, *Batteries*, De la Peña, *Body Electric*, Nye, *Electrifying*, and Williams, "Technology and Media."
8. Schaefer, "Ethernet."
9. Cohen, *A Body Worth Defending*.
10. Moreover, even these distinctions, Bruno Latour argues in *We Have Never Been Modern*, are illusory and more rhetorical than actual, even in "modernity."
11. Milutis, *Ether*.
12. Genesis 2:7, King James Version.
13. E.g., *Them*, *The Beast from 20,000 Fathoms*, *The Beginning of the End*, *Godzilla*, *The Deadly Mantis*, *It Came from Beneath the Sea*, *Tarantula*.
14. Douglas, *Listening In*, Milutis, *Ether*, Sconce, *Haunted Media*, Simon, *Dark Light*, and Sterne, *Audible Past*.
15. Sidney, "Whistling," "Training."
16. *1408*, *Beyond Darkness*, *The Caller*, *Disconnect*, *White Noise*, *White Noise 2*, *Shutter* (US and Thai versions), *Frequency*, *Poltergeist*, *The Ring* films, and *Photographing Faeries*. Variants of this supernatural connection include *Electrick Children*, in which a woman believes God impregnated her through an audio cassette tape player. Extra-material connections can

also be extraterrestrial crossovers, as when a television connects to a ravenous alien life form in *TerrorVision* or virtual pet toys are the conduit for brainwashing aliens in "The Tale of the Virtual Pets" episode of *Are You Afraid of the Dark?*

17. Foucault, *Abnormal*, 223.
18. Wagner, "Hellish Phone Call"; Dikeman, "Terrified by Ghost"; Herz, "Randy Quaid"; Walker, "Ida Lupino."
19. Supernatural connections, it should be noted, can also be benign, although that lies outside the scope of this project. In *The Manitou*, technologies are not the conduit for supernatural possession but for exorcism, demonstrating a beneficent version of techno-supernatural-biological entanglement. In this 1978 film, Susan Strasberg plays a woman possessed by a 400-year-old, evil, Native American medicine man, who afflicts her with a tumor-like dermoid fetus growing on her back. Tony Curtis plays a San Francisco tarot card reader who discovers that "machines have their own *manitou*" or universal spirit energy. In the film's conclusion, Strasberg's character channels the spirit-energies of a room full of computers and electric medical equipment to defeat the evil spirit, presenting a white, diseased, mother-medium triumphant against her native attacker. In the novel, the evil spirit is stopped by venereal disease.
20. Franklin, *Experiments and Observations*, 117.
21. "Animal" in this usage is inclusive of, not opposite to, humanity, as in the categorical triad of animal, vegetable, or mineral.
22. On electricity, biology, physics, ether, and related histories, see Asendorf, *Batteries*, Ashcroft, *Spark*, Beard, *Salem Witchcraft*, Bichat, *Physiological Researches*, Bloch, *Mesmerism*, Brazier, *Electrical Activity*, Cantor and Hodge, *Conceptions*, Craig, *Influence of Variations*, Darnton, *Mesmerism*, De Laurence, "*Magnetic Hypnotism*", De la Peña, *Body Electric*, Dibner, *Galvani-Volta*, Fara, *Entertainment for Angels*, Fuller, *Mesmerism*, Faraday, *Progression*, Franklin, *Experiments and Observations*, Fraser, *Medical Electricity*, Gale, *Electricity*, Green, *Galvani on Electricity*, Grimes, *Etherology*, Haigh, *Xavier Bichat*, Hales, *Statical Essays*, Heilbron, *Electricity*, Heydenfeldt, *Conscious Forces*, Home, *Electricity*, Marvin, *Old Technologies*, Metzger, *Blood and Volts*, Milutis, *Ether*, Morus, *Frankenstein's Children*, Nye, *Electrifying*, Peters, "Broadcasting and Schizophrenia", Putnam, *Mesmerism*, Ochs, *Nerve Functions*, Reichenbach, *Odic-Magnetic Letters*, Rockwell and Beard, *Clinical Researches*, *Practical Treatise*, Schiffer, *Draw the Lightning*, *Power Struggles*, Simon, *Dark Light*, Snow, *Matter and Gravity*, Thomson, *Light Thrown*, Waterhouse, *Principle*, Wells, *Theory of Disease*, and Wright, *Human Existence*.
23. Hutchison, "Occult," 233.
24. Heilbron, *Electricity*.
25. Henry, "Occult," 338.
26. Craig, *Influences of Variations*.
27. Ibid. 24.
28. Ibid. 25.
29. Ibid. 25.
30. Ibid. 26.
31. Ibid. 28.
32. Rockwell and Beard, *Practical Treatise*, 290, emphasis mine.

33. Craig, *Influence of Variations*, 9.
34. Ibid. 254–255.
35. Wells, *Theory of Disease*, 21–22, 36.
36. Shelly, *Frankenstein*, Chapter 5, ¶1.
37. Beard, *Salem Witchcraft*, 28.
38. Ibid. 28.
39. Milutis, *Ether*.
40. Sconce, *Haunted Media*.
41. Becker and Selden, *Body Electric*.
42. Tercier, "Lips of the Dead."
43. Kennedy, *History*.
44. Hutcheon and Hutcheon, *Opera*; Kennedy, *History*.
45. Simon, *Dark Light*, 150, emphasis orig.
46. "Govt. Cancels," Kuper, "Radiation Hazards."
47. Burgess, *Cellular Phones*, 98.

References

1408. Dir. M. D. Håfström. 2007, USA.

Asendorf, Christopher. *Batteries of Life: On the History of Things and their Perception in Modernity*, University of California Press, 1993.

Ashcroft, Frances. *The Spark of Life: Electricity in the Human Body*, WW Norton & Company, 2012.

Associated Press. "ER Docs: Don't Text and Walk, Skate—or Cook." *USA Today*, 2008, July 30.

Beard, George Miller. *The Psychology of the Salem Witchcraft Excitement of 1692 and Its Practical Application to Our Own Time*, G. P. Putnam's Sons, 1882.

Becker, Robert, and Gary Selden. *The Body Electric: Electromagnetism and the Foundation of Life*, Morrow, 1985.

Bellett, Gerry. "Shocking News about Earbuds and Lightning." *Canada.com*, 2007, July 12.

Beyond Darkness. Dir. C. Fragasso, C. aka Clyde Anderson. 1992, Italy.

Bichat, Xavier. *Physiological Researches on Life and Death*, Richardson and Lord, 1827.

Bloch, George, ed. *Mesmerism: A Translation of the Original Scientific and Medical Writings of F. A. Mesmer*, William Kaufmann, Inc., 1980.

Brazier, Mary A. B. *A History of the Electrical Activity of the Brain: The First Half-century*, Pitman Medical Publishing, 1961.

Burgess, Anthony. *Cellular Phones, Public Fears and a Culture of Precaution*, Cambridge University Press, 2004.

The Caller. Dir. Matthew Parkhill. 2011, USA.

Cantor, G. N., and M. J. S. Hodge, eds. *Conceptions of Ether: Studies in the History of Ether Theories 1740–1900*, Cambridge University Press, 1981.

Cohen, Ed. *A Body Worth Defending: Immunity, Biopolitics, and the Apotheosis of the Modern Body*, Duke, 2009.

"Could an iPod Become a 'Frypod'?" *Inside Edition*. TV broadcast 2006, July 16. http://www.insideedition.com/ourstories/inside_stories/story.aspx?storyid=854.

Craig, William. *On the Influence of Variations of Electric Tension as the Remote Cause of Epidemic and Other Disease*, J. Churchill, 1859.

Darnton, Robert. *Mesmerism and the End of the Enlightenment in France*, Harvard University Press, 1968.

De la Peña, Carolyn. *The Body Electric: How Strange Machines Built the Modern American*, NYU Press, 2005.

De Laurence, L. W. "*Magnetic Hypnotism*," "*Medical Hypnosis*," and "*Suggestive Therapeutics*," de Laurence, Scott and Co., 1910.

Dibner, Bern. *Galvani-Volta: A Controversy that Led to the Discovery of Useful Electricity*, Burndy Library, 1952.

Dikeman, M. "We Were Terrified by Ghost who Spoke Through TV Set." *National Enquirer*, 1982, January 5, p. 60.

Disconnect. Dir. Robin Christian. 2010, USA.

Douglas, Susan J. *Listening In: Radio and the American Imagination*, Random House, 1999.

Electrick Children. Dir. Rebecca Thomas. 2012, USA.

Fara, Patricia. *An Entertainment for Angels: Electricity in the Enlightenment*, Columbia University Press, 2002.

Faraday, Michael (attributed through spirit medium). *Progression, or How a Spirit Advances in Spirit Life, the Evolution of Man. Two Papers, Given in the Interest of Spiritual Science, by Michael Faraday, late Electrician and Chemist of the Royal Institution, London, England*, Star Publishing, 1885.

Fitz, R. "The Shocking Truth about iPods." *National Enquirer*, 2007, August 6, p. 46.

Foucault, Michel. *Abnormal: Lectures at the Collège de France, 1974–1975*, Picador, 2004.

Franklin, Benjamin. *Experiments and Observations on Electricity, Made at Philadelphia in America, to which are Added, Letters and Papers on Philosophical Subjects*, David Henry, 1769.

Fraser, E. J. *Medical Electricity: A Treatise on the Nature of Vital Electricity in Heath and Disease*, C. S. Halsey, 1863.

Frequency. Dir. Gregory Hoblit. 2000, USA.

Fuller, Robert C. *Mesmerism and the American Cure of Souls*. Philadelphia: University of Pennsylvania Press, 1982.

Gale, T. *Electricity, or Ethereal Fire, Considered: 1st, Naturally, as the Agent of Animal and Vegetable Life; 2d, Astronomically, or as the agent of Gravitation and Motion; 3d, Medically, or its Artificial use in Diseases. Comprehending both the Theory and Practice of Medical Electricity; and Demonstrated to be an Infallible cure for Fever, Inflammation, and Many Other Diseases: Constituting the Best Family Physician Ever Extant*, Moffitt and Lyon, 1802.

"Govt. Cancels Warning About X-Ray Danger from Watching Color TV Sets," *National Enquirer*, 1972, January 9, p. 23.

Green, Robert Montraville, ed. *Galvani on Electricity*, Elizabeth Licht, 1953.

Grimes, J. Stanley. *Etherology; or the Philosophy of Mesmerism and Phrenology; including a New Philosophy of Sleep and of Consciousness, with a Review of the Ppretensions of Neurology and Phreno-magnetism*, Saxton and Miles, 1845.

Haigh, Elizabeth. *Xavier Bichat and the Medical Theory of the Eighteenth Century (Medical History, Supplement No. 4)*, Wellcome Institute for the History of Medicine, 1984.

Heffernan, E. J. "Thunderstorms and iPods—Not a Good iDea." *New England Journal of Medicine*, vol. 357, no. 2, 2007, pp. 198–199.

Heilbron, J. L. *Electricity in the 17th and 18th Centuries: A Study of Early Modern Physics*, University of California Press, 1979.

Henry, John. "Occult Qualities and the Experimental Philosophy: Active Principles in Pre-Newtonian Matter Theory." *History of Science*, vol. 24, 1986, pp. 335–381.

Herz, S. "Randy Quaid: My Dead Dad Caused a Power Outage." *National Enquirer*, 1996, July 30, p. 40.

Heydenfeldt Jr., S. *The Unison of the Conscious Forces (continued)*, W. M. Hinton and Co., 1890.

Home, W. *Electricity and Experimental Physics in Eighteenth-century Europe*, Variorum/Ashgate, 1992.

Hutcheon, Linda, and Hutcheon, Michael. *Opera: Desire, Disease, Death*, University of Nebraska Press, 1996.

Hutchison, Keith. "What Happened to Occult Qualities in the Scientific Revolution?" *Isis*, 1982, pp. 233–253.

Kennedy, Michael. *A Brief History of Disease, Science and Medicine from the Ice Age to the Genome Project*, Asklepiad Press, 2004.

Kuper, J. B. Horner. "Radiation Hazards." Letters to the Editor of The Times. *New York Times*, 1969, January 5, E13. Proquest database. Accessed June 27, 2007.

Latour, Bruno. *We Have Never Been Modern*, Harvard University Press, 1991.

The Manitou. Dir. William Girdler. 1978, USA.

Marvin, Carolyn. *When Old Technologies were New: Thinking about Electric Communication in the Late Nineteenth Century*, Oxford University Press, 1988.

Matyszczyk, Chris. "Man Killed by Lightning Because of Selfie Stick?" *CNET*, 2015, July 6. https://www.cnet.com/news/man-killed-by-lightning-because-of-selfie-stick/. Accessed July 16, 2015.

Metzger, Tom H. *Blood and Volts: Edison, Tesla, and the Electric Chair*, Autonomedia, 1996.

Milutis, Joe. *Ether: The Nothing that Connects Everything*, University of Minnesota Press, 2006.

Morus, Iwan Rhys. *Frankenstein's Children: Electricity, Exhibition, and Experiment in Early-Eighteenth-Century London*, Princeton University Press, 1998.

Nye, David E. *Electrifying America: Social Meanings of a New Technology, 1880–1940*, The MIT Press, 1992.

Ochs, Sidney. *A History of Nerve Functions: From Animal Spirits to Molecular Mechanisms*, Cambridge University Press, 2004.

Peters, John Durham. "Broadcasting and Schizophrenia." *Media, Culture and Society*, vol. 32, 2010, pp. 123–140.

Photographing Faeries. Dir. Nick Willing. 1997, UK.

Poltergeist. Dir. Tobe Hooper. 1982, USA.

Putnam, Allen. *Mesmerism, Spiritualism, Witchcraft, and Miracle*, John Wilson and Son, 1858.

Reichenbach, Baron Charles aka Carl Freiherr von Reichenbach. *Odic-Magnetic Letters*, Calvin Blanchard, 1860.

The Ring. Dir. Gore D. Verbinski. 2002, USA.

Ring 0: Birthday. Dir. T. Norio. 2000, Japan.

The Ring Two. Dir. Hideo Nakata. 2005, USA.

The Ring Virus. Dir. D. Kim. 1999, South Korea.

Rings. Dir. J. D. Liebesman. Motion picture included as bonus feature on DVD release of *The Ring Two*. 2005, USA.

Ringu. Dir. Hideo Nakata. 1998, Japan.

Ringu 2. Dir. Hideo Nakata. 1999, Japan.

Rockwell, A. D., and George Miller Beard. *A Practical Treatise on the Medical and Surgical uses of Electricity, Including Localized and General Electrization*, William Wood and Co., 1871.

Rockwell, A. D., and George Miller Beard. *Clinical Researches in Electro-surgery*, William Wood and Co., 1873.

Ryan, J. "Warning: Be on the Look Out for Killer iPods!" *The Post Chronicle*, 2007. http://www.postchronicle.com/news/original/article_21262969.shtml. Accessed February 7, 2007.

Schaefer, Pete. "Why is 'Ether' in 'Ethernet?" *International Journal of Communication*, vol. 7, 2013, pp. 2010–2026.

Schiffer, Michael Brian. *Draw the Lightning Down: Benjamin Franklin and Electrical Technology in the Age of Enlightenment*, University of California Press, 2003.

Schiffer, Michael Brian. *Power Struggles: Scientific Authority and the Creation of Practical Electricity Before Edison*, The MIT Press, 2008.

Sconce, Jeffrey. *Haunted Media: Electronic Presence from Telegraphy to Television*, Duke University Press, 2000.

Shelly, Mary Wollstonecraft. *Frankenstein, or The Modern Prometheus*. *Literature.org*. http://www.literature.org/authors/shelley-mary/frankenstein/. Accessed June 8, 2013.

Shutter. Dirs. Banjong Pisanthanakun, Parkpoom Wongpoom. Thailand, 2004.

Shutter. Dir. M. Ochiai. 2008, USA.

Sidney, Frederic H. "The Whistling Telegraph." *Overland Monthly*, 1919, June, pp. 536–540.

Sidney, Frederic H. "Training a Spirit to Telegraph" *Overland Monthly*, 1919, June, pp. 131–134.

Simon, Linda. *Dark Light: Electricity and Anxiety from the Telegraph to the X-ray*, Harcourt, 2004.

Snow, A. J. *Matter and Gravity in Newton's Physical Philosophy*. London: Oxford University Press, 1926.

Sterne, Jonathan. *The Audible Past: Cultural Origins of Sound Reproduction*, Duke University Press, 2003.

"The Tale of the Virtual Pets." Episode of *Are You Afraid of the Dark?* Dir. Iain Paterson. 1999, USA.

Tercier, J. "The Lips of the Dead and the 'Kiss of Life': The Contemporary Deathbed and the Aesthetic of CPR." *Journal of Historical Sociology*, vol. 15, no. 3, 2002, pp. 283–327.

TerrorVision. Dir. Ted Nicolaou. 1986, USA.

Thomson, J. J. *On the Light Thrown by Recent Investigations on Electricity on the Relation between Matter and Ether*. Manchester University lectures: The Adamson Lecture delivered at the University on November 4, 1907, Manchester University Press, 1908.

Tomes, N. "Epidemic Entertainments: Disease and Popular Culture in Early-twentieth-century America." *American Literary History*, vol. 14, no. 4, 2002, pp. 625–652.

Wagner, D. "Hellish Phone Call." *Weekly World News*, 2007, July 30, p. 11.
Walker, D. "Ida Lupino Tells of Phone Call from a Dead Friend." *National Enquirer*, 1969, July 20, p. 3.
Waterhouse, Benjamin. *On the Principle of Vitality*, Thomas and John Fleet, 1790.
Wells, W. R. *A New Theory of Disease; Based upon the Principle that Man is a Compound Electrical Magnet*, Tracy and Rew, Evening Express Office, 1869.
White Noise. Dir. G. Sax. 2005, USA.
White Noise 2: The Light. Dir. Patrick Lussier. 2007, USA.
Williams, Rosalind. "Technology and Media." Paper presented at the University of Southern California, Visual Studies Graduate Certificate program, 2009, March 6.
Wright, Thomas L. *Notes on the Theory of Human Existence, Comprising Remarks on Vitality, the Mind, and Incidentally, the Soul, the Whole Being an Exposition of the Nature, Powers, and Destiny of Man*, Collins and Van Wagner, 1848.

· 4 ·

THE ELECTRICAL BANAL

Anderson, SC, "The Electric City"

Technology historian David Nye writes that during the period of electrification in the United States, "the self and the electrical world have intertwined."[1] However, in the previous chapter, I have argued that the self and electricity were always already entangled, and what we see instead historically are processes of separation and thingification, producing new and what appear to be distinct entities, bodies of knowledge, and practices that were once unified or overlapping. Separation and thingification make technopathologies more conceivable by allowing for the cause-and-effect relationship of disease, but, as I will now demonstrate, not necessary.

To dismiss technopathological discourses as solely technophobic reactions to new technologies supports the ideology of new technologies typically provoking utopian and dystopian reactions. In this chapter, I will show that this is not always the case. At the turn of the twentieth century, "The Electric City," Anderson, South Carolina, was a regional, national, and global leader in the new technologies of electrification. It was a situation of extreme technological newness, but, contrary to common narratives of technological history, residents of the Electric City typically did not have utopian or dystopian reactions. Instead, I will show how electricity was rationally integrated into disparate dimensions of individual and social life. Instead of viewing technological emergence and adoption in terms of utopian and dystopian reactions (with technopathologies misunderstood as a component of the latter), using an agential realist approach to the ongoing processes of becoming users finds a fairly sober and pragmatic integration of electricity into multiple dimensions of life. In contrast to theorizations of an "electrical sublime," here we see what I refer to as an "electrical banal." In media discourses on electricity from Anderson and surrounding areas, we see, rather than a disruptive technological impact, many ways in which the new flows of electricity were accepted into the lives, economies, and bodies of users in a most quotidian fashion.

If the standard view is that reactions to new technology tend to be utopian or dystopian, we should encounter as much in a town that was an early adopter of electricity. Especially in semi-rural South Carolina, where technological progress was associated with escaping from the past and emerging as a "New South." Instead, in Anderson there was decided ambivalence. Some reactions were negative; others were positive, but most were a sober integration into daily life, rather than hyperbolic reaction to something new. This suggests that electricity was not viewed in a deterministic manner, as some strange external force changing their town, but as something complex and closer to human bodies and nature. Ambivalence in Anderson supports my argument of seeing technopathologies as logical, with longstanding roots, as previously discussed.

Foucault writes, "How light power would be, and easy to dismantle no doubt, if all it did was to observe, spy, detect, prohibit, and punish; but it incites, provokes, produces. It is not simply eye and ear: it makes people act and speak."[2] In this section I explore the productive power of emergent electrical power in the case study of Anderson. As a regional and national leader in electrification, what actions and utterances did the rearrangement of power relations and the creation of power networks incite? What bodies and subjects were (co)produced as electricity was domesticated? What entanglements appear in the phenomena that are (re)becoming?

Topianism and the Electrical Sublime

The common view of emergent technologies is one of reaction: They cause various effects in citizens, often in hyperbolic extremes of utopian and dystopian visions. Several scholars have described such anxious, spectacular, and supernatural associations with electrical technologies, such that it has become conventional wisdom of technological history.[3] For example, in the late 1800s, there were concerns over early electric lighting causing blindness and "photo-electric ophthalmia"[4] or "electric light eye."[5] There were fears that public nighttime illumination would lead to increasing drunkenness and degenerate behavior, aiding thieves, enabling bar-crawling, and releasing prostitutes from brothels into the streets.[6] Artificial illumination did not make people drunk or criminal, but it made possible greater opportunities for their sicknesses.

I question the degree to which this notion has become conventional wisdom—or perhaps ideology—in technology studies. For one, its framing of events in terms of cause-and-effect, or stimulus-and-reaction, is both technologically determinist and pathologizing, when one considers the conception of pathology as a reaction to one's environment, albeit inferior. Second, in focusing on individual reactions, it neglects social productions.

The "electrical sublime" is historian David Nye's evolution of his broader concept of American "technological sublime." According to Nye, the awe-struck sense of simultaneous wonder and fear that was once reserved for sacred, religious experiences became transferred to the natural wonders of the American continent as it was explored and settled. This sense of the sublime then shifted to technologies that conquered nature, such as the engineering marvels of dams and bridges. The sublime then spread to that which powered such technologies. The "electrical sublime" applied to that which conquered, not only nature, but, as previously discussed, supernature. Spectacular public displays of illumination celebrated human mastery over nature as quasi-magical, almost religious events. The public venerated electricity as a magical cure-all and etheric bridge between the mind, body, spirit, and sexuality, and the realms of the natural and supernatural worlds. Carolyn Marvin supported this view as well, arguing that a key dimension of this early "electrical imagination" was bending electrical discourses to the familiarity and authority of the body. "Such was the cogency of bodily authority that even sophisticated observers entertained magical convictions about the relationship between electricity and their bodies," such as electric adornments and therapies, freaks with electrical powers, sickness induced by electricity, and cures effected by it.[7] Subsequent technology scholars have expanded on this concept, such as Vincent Mosco's description of a "digital sublime."

This techno-historic narrative of the sublime and -topian reactions has become common in cultural histories of technology. As such, it begins to fall into the realm of "of course, everybody knows that"—the ideological. In this chapter, I explore a counter-history of electrification to attempt to unpack some of the ideological work of the idea of the sublime. Topianism imposes a dualistic worldview. It maintains cause and effect, technological determinism, and an ontology of separate things rather than entangled processes, and emphasizes reaction over production. Most obviously, it pathologizes users in their irrational reactions while simultaneously normalizing them: This is how people always react to new technology. This undermines careful examination of actual social processes, and introduces a smug sense of presentism, even when recent examples are pointed out. Topian perspectives reinforce users as stable subjects, as coherent things upon whom technology has effects, which neglects poststructural and queer insights into the instability and variability of subjects and phenomena. It is deterministic, focusing on technology and neglecting issues of human agency. Even if the causal direction is reversed or made two-way, it is still a story of separate things engaged in cause and effect. Such separation between humanity and technology is a false dualism, like that of nature and culture. It is also synchronic. That is, it creates a snapshot of events and moments: a reaction, a change, an effect. While important inquires, such a focus on moments of things interacting comes at the expense of diachronic, longitudinal explorations, and inquiries about ongoing, entangled processes and phenomena.

Mobile technology scholar Cara Wallis adapted Stuart Hall's essay title, "Marxism without Guarantees" as "mobiles without guarantees," meaning the effects and impacts of mobile technologies on different cultures cannot be universalized or generalized like natural law. I expand this more to argue that Anderson suggests *newness without guarantees*.

A risk of topian narratives is their reinforcement of a historic disjuncture, such as Kittler's romantic/modern discourse networks. While not disputing processes of historical change, there is danger of a modern nostalgia for romanticism. The old and new are not separate, but, like Self and Other or East and West, involved in mutual constitution, entanglement, connection, and contagion. It is a cliché that stories of the past are always about the time in which they are written. Stories of topian reactions to new technologies in the past risk reassuring us that were are removed from such reactions, evolved beyond them. Rather than perform the critical work of denaturalizing or queering the present, this serves to reinforce the status quo.

In order to investigate Anderson's electrical history, I visited the town several times to explore local monuments and historic markers. Focusing on the period of 1890–1920, I researched at the county library and history museum, examining archives of newspapers, documents, oral history transcripts, industry publications, local papers, clippings, company publications and scrapbooks, and local independent historiographies. These incomplete and varied collections were augmented with online archives of Anderson and other South Carolina newspapers through the Library of Congress Chronicling America database. In drawing on multiple newspapers—the main paper, the *Anderson Daily Intelligencer*, a more agrarian and populist competitor, the *Anderson Peoples Advocate*, and papers from other towns—I aimed for a variety of sources to allow for differing motives of editors and publishers, yet the results were fairly consistent.

What we see in Anderson is not limited to atomistic individuals reacting to technological things—electricity, illumination, telephony. Instead, we see a clear example of a complex process of integrating technological phenomena into multiple facets of daily lives. Andersonians were becoming consumers of electricity, but also users of it, as they applied it to different tasks, including identity work and social projects. Electricity did not make them into modern citizens of the New South; they used it in their own attempts to become such. In ongoing processes, Anderson, SC, becomes and re-becomes The Electric City. The residents become electrical subjects and often use the new technologies in ways that reinforce other social phenomena. Although electricity is new, a perspective of its integration—rather than reactions to it—help show how it was entangled with ongoing becomings of race, gender, and power hierarchies.

"The Electric City" Today

Anderson sits in the northern, Upstate region of South Carolina, a largely semi-rural and agricultural area previously dominated by the textile industry. Located just off Interstate 85, it sits almost equidistant between Atlanta, Georgia, and Charlotte, North Carolina, each about 120 miles away. With a city population of 27,000, it houses primarily automotive and plastics manufacturing, as well as Anderson University, a private Christian school. Like most towns in this region, it features an outer region of commercial sprawl and unrestrained development of chain restaurants, discount stores, and strip malls, but a historic downtown in the old city center.

Anderson was dubbed "The Electric City" in 1895 and has continued to use the appellation into the 1930s and today. The community newspaper is *The Electric City News*, and the official visitor's guide to downtown Anderson pictures a hydroelectric generator wheel and statue of its developer, along with the tagline "spark your imagination." The brochure urges visitors to "Electrify Yourself!" in a town that "shines and delights."

Downtown's Main Street features a statue of native son and electrical innovator William Church Whitner in front of the red-brick courthouse. A block off Main Street, on Whitner Avenue, nestled between parking lots on a street corner across from a restored 1939 theater, sits Generator Park, a "pocket park." It features 10,000 square feet of manicured grass, red brick paths and steps, and small landscaped shrubs and trees. At its center, the paths converge around a massive, black, hydroelectric generator wheel, 13 feet in diameter and weighing 50,000 pounds, mounted atop four cement columns and looming over visitors. It affords a stunningly tactile, concrete encounter with mechanical engineering technology. Yet, as one reads the historical markers on the cement supports, or walks up the steps to the state marker, the generator wheel becomes dwarfed itself by a row of two-story brick storefronts, including a former hardware store, now empty and shuttered. A limp real estate banner sags loosely from the front of the center building (Fig. 4.1).

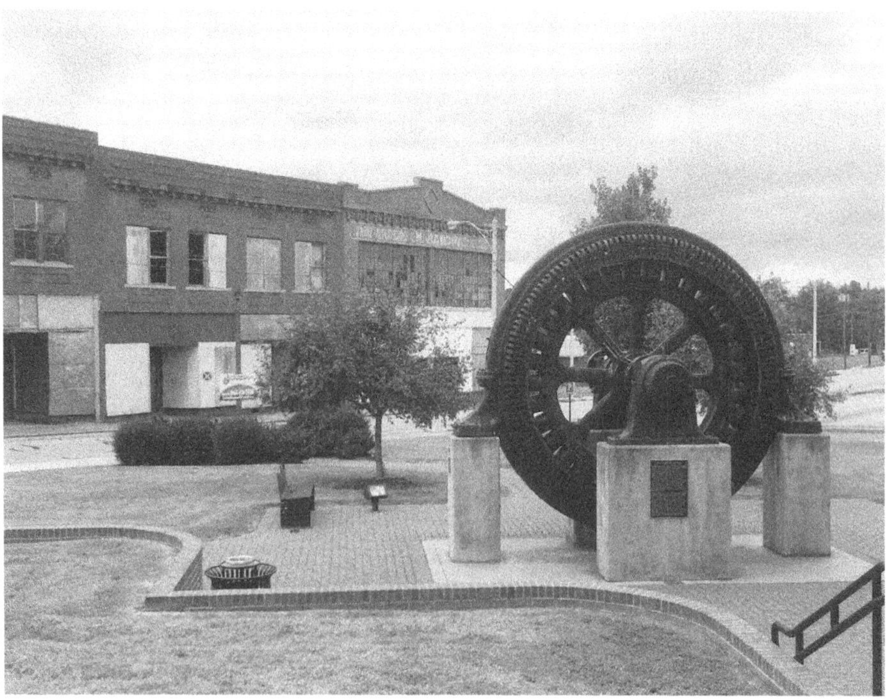

Figure 4.1. Generator Park, downtown Anderson, SC. Park Shoals hydroelectric generator wheel in foreground with empty buildings in background, 2014.
Source: Photo by the author.

Old and new co-exist in Anderson, whether they are modern innovation and antebellum nostalgia, or abandoned buildings and new manufacturing jobs with European automakers, which moved here for cheap labor, minimal regulations, and lack of unionization. More importantly, old and new also co-existed at the time of electrification.

Rebecoming Electrical

The term "electrification" oversimplifies the commercial implementation and adoption of electrical power systems, which was a complex process involving numerous actors and institutions operating with varied rates and manners.[8] Rather than risk such generalizations, I will limit my discussion primarily here to my case study of Anderson and aspects of wider electrical history insofar as they relate to Anderson.[9]

As has been well documented, a central conflict during the years of electrification was what has been called the "current wars," a battle for adoption and lucrative contracts between Thomas Edison's technology of direct current (DC) and George Westinghouse's alternating current (AC). A distinction between the two was that, although AC was ostensibly more powerful, DC was considered safer and could be more easily transmitted over long distances.

Electrification was an uneven process, but, in many ways, Anderson was a leader. In 1897, during these "current wars," Anderson became the first city in the nation, and possibly the world, to send high-voltage AC electricity long distances without using transformers (devices that change alternating current from one voltage to another). AC was considered very dangerous, and heretofore mills used steam power off of adjacent rivers and streams. This success led to numerous claims of Anderson being first, continuing to today. It was the first electric-powered cotton mill in the nation. Anderson was the first city in the United States to have a continuous supply of electric power.[10] It had the first textile mill in the United States with long-distance power, from the only AC plant in the world with a voltage as high as 11,000. This plant powered the world's first electric cotton gin, developed by Andersonian Oliver Bolt. Anderson was the first city in South whose main industry was powered by electricity. By 1911, South Carolina had become the first state with long-distance electrical power lines, with double the miles of high voltage lines of any other state except New York.

William Whitner had begun working in railroad engineering but succumbed to typhoid fever and had an extended convalescence at his father's home in Anderson. While there, the city asked the 26-year-old man to build a water works system and steam electric plant. In 1888, when the charter was approved, Anderson had a population 3000. Whitner completed a steam-powered electric plant in 1890, the Anderson Waterworks, and installed 750 incandescent lamps using steam-powered electricity, but the plant proved too expensive to operate. Inspired by European experiments with long-distance water power, he contemplated generating power from the river currents and delivering it by wire. He traveled to New York to seek advice from Nikola Tesla in 1891, then returned to Anderson. Because of the current wars, Andersonian colleagues had doubts about alternating current, and it took much convincing to raise capital, inspiring jokes and skepticism in town. Transmitting high-voltage power over long distances without transformers was considered too risky; only low voltages could be transmitted safely, and for short distances.

In 1893, Whitner convinced backers and the directors of Anderson Electric Light and Power Company that high-voltage power should and could be used. Whitner leased the High Shoals plant on the Rocky River in 1894 and, in an experimental demonstration, proved Tesla correct: high-voltage alternating current could be transmitted over a distance. On May 1, 1895, the 5,000-volt AC generator at McFall's Mill at High Shoals transmitted power six miles to water system pumps at the Tribble Street power and water yard in Anderson. Success enabled him to raise funds to build a larger dam and power plant at Portman Shoals on the Seneca River, 11 miles west of Anderson. The Portman Shoals plant went into operation in 1897, providing electricity for the town, its arc lights, its first telephone exchange, and pumping water to the waterworks.

On June 19, 1895 the city's main paper, the *Daily Intelligencer*, ran a front-page story headlined, "The Electric City."

> We have the proud satisfaction of announcing to our readers that this number of the *Intelligencer* is issued from a press in our office run by electricity, and that thereafter we expect to use this power in the publication of our paper. Can any of our contemporaries in the State or in the South say as much? … We feel very proud to be able to boast of having the pioneer Company in the State, or indeed in the South, as we are informed, to demonstrate the practicality of electric power by what is known as the long-distance transmission system; and we hope that our contemporaries will take due notice of the fact and give to us our due need of praise.
>
> We take particular pleasure in calling the attention of the [Charleston] *News and Courier*, that dubbed us "The Electric City," to our achievement in this respect, and

> to remind it that here is another step in that industrial progress in the State in which it has become the acknowledged champion and exponent.[11]

In addition to thanking specific parties involved, the article encouraged other businesses to come to Anderson to take advantage of its electrical supply. They did. Already a cotton boomtown, the population of Anderson had doubled between 1890 and 1895. Now, with cheap electricity, it continued to grow rapidly. The plant was upgraded for greater power capacity.[12] In October of 1897, the Electric City Shoe Store opened.

Anderson, as a model for the nation, found its size and reputation continued to grow into the new century. President Taft visited in 1911, and in 1914 the city's rapid growth was forcing the hotel to use cots in its lobby for the overflow of visitors. At this time, the *Daily Intelligencer* ran columns called "Electric City Sparklets" and "Electric Epigrams," an article on "Electric City Pride," often referred to outstanding citizens as "live wires," and advertised movies at The Electric Theatre. In 1899, the city had claimed 96 light customers, 17 users of power, 73 arc lamps, and four 32-candlepower lamps lighting the city. By 1924, this had grown to 2,598 lighting customers, 188 power users, and streets had 225 400-candlepower lamps, 41 60-candlepower lamps, and 580 lamps on 116 white-way posts.

Whitner went on to create hydroelectric power stations for towns across the South. Under a succession of owners, Portman Shoals continued producing electricity until the 1950s, when the area was dammed and flooded to create Lake Hartwell.

Ingredients for Topianism

As "The Electric City," suffused in technological newness, Anderson should match or exceed the utopian and dystopian reactions described in technology histories. Anderson had all the ingredients for topian reactions—but doesn't evidence having them. I will now describe the elements present during Anderson's electrification, which exhibit not only the roots of pathological technoculture previously discussed, but also the general makings for topian reactions. Instead, however, I will show how Andersonians integrated electricity—rather than reacting to it. Electricity, already an entangled part of human bodies, minds, and souls, does not enter into culture with electrification but becomes instead reoriented and reframed into new aspects of daily life as a commodity—the electrical banal in action.

As I have described, "The Electric City" was engulfed in newness. If technopathologies were reactions to newness, they should be more evident, or at least other topian reactions. Not only was Anderson on pace or ahead of the process of electrification in general, it was especially ahead of the curve for non-urban areas. Rural areas lagged behind the rest of the country in electrification. Brox has described how electrification intensified resentment between electrified cities and unelectrified rural areas, who discovered what they were missing through media reports and product catalogs.[13] This association of electrification with urbanism and modernity should have made it all the more disruptive in Anderson, which was not only a rural textile community, but also located in South Carolina. Only a few decades after the end of the Civil War, Southern states such as South Carolina were particularly invested in notions of modernity and technological progress as an imagined route to returned prosperity and prominence for the region, and a way to separate from the past of slavery, war, and Reconstruction. As I will describe later, this "New South" was part of what was produced during electrification.

Electricity and its technologies were spectacular. Anderson first saw electric lights in a traveling circus in 1880. In 1910, the circus brought elephants using the telephone (Fig. 4.2). By that year, Anderson had its own "white way" downtown, decorated with thousands or white and colored lights, and newspapers reported on cities in other states that had acquired similar magnificent attractions.

During this period, Andersonians exhibited evidence of the conflated, universal theories of equivalence that connected biological, supernatural, and mechanical electricities I discussed in a previous chapter. The establishment of the mill in the 1890s occurred during a time of millennialism, when religious tracts in the area prophesized "the most astounding and constantly augmenting manifestations of the invisible spiritual powers, both good and evil."[14] The spiritualism of theosophy was discussed in the *Anderson People's Advocate*.[15] The mill was described as using "'that imponderable and invisible agent, long distance electricity'" or "'white fire'."[16] A more colorful example appeared in an Anderson newspaper's reprint of an Indiana story warning of the dangers of lightning bugs (fireflies):

> The medical journals have been calling attention to the danger that may result from handling the electric light bugs, and make the statement that they are very venomous. Several cases of severe poisoning have been reported from their bites, and it will be well to let them alone. A great many children play with them under the impression that they are harmless, but they ought to be warned against the practice. These

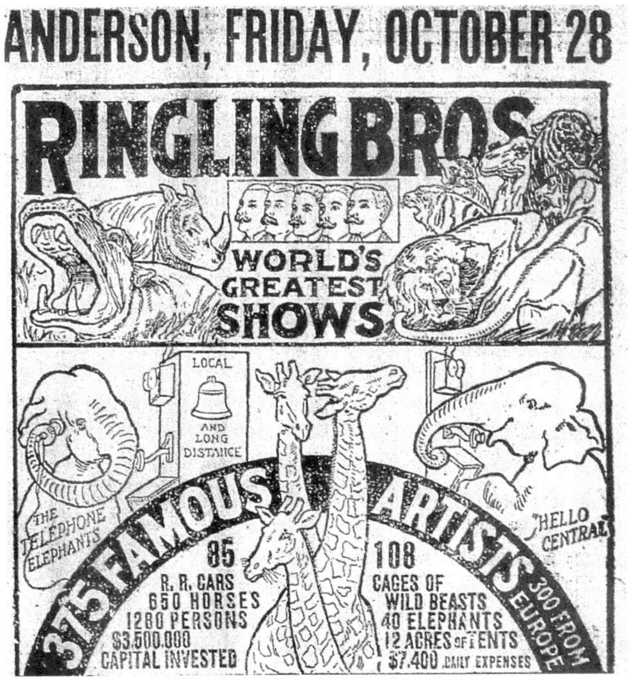

Figure 4.2. Spectacles of new technology: elephants using the telephone. Advertisement for Ringling Brothers circus. Anderson, SC.
Source: *Anderson Peoples Advocate* (1910, October 10): 3.

> bugs never made their appearance until the advent of the electric lights, and they are spread over the whole country.[17]

The entanglement of biological and mechanical energies appears also in references to electrical health. In the early 1890s, the *Intelligencer* and the *Advocate* ran ads for "Electric Bitters," a tonic promoted here and in other states for curing flu, bad digestion, and "female trouble." Flu was called "the demon of the air," referencing earlier miasmic theories of disease.[18] Other stories at this time reported on French doctors "electroplating human bodies" to preserve corpses, disease being caused by, not personal susceptibility, but a lowering or weakening of "life forces" or "vital resistance," and using "mind cures" to heal rheumatism and paralysis through an outside stimulus, after which "new and powerful nervous currents are set in motion, and these currents are sent by the brain to the diseased parts. They are currents that carry strong, bounding life."[19]

Finally, there is also evidence of early suggestions of pathologizing media content and technologies. The violent content of motion pictures was feared to provoke real-life violence.[20] Nocturnal outdoor lighting was aligned with degenerate behavior.[21] Telephone and telegraph wires were reported to attract and transmit lightning to users.[22]

Not So Sublime

And yet, overall, Anderson had very few extreme reactions. Other than initial skepticism toward Whitner's first experiment, there was no apparent controversy, no dystopian fear of electrical nightmares, no life-threatening sublime power of illumination or telephony. The concerns about electrification were described as just that: rational, pragmatic concerns; and they were won over. Even the *Advocate*, which began as a progressive publication for farmers, but, by the 1910s, had become a sensationalistic paper running stories of murder, suicide, and cannibalism, only reported minimal coverage of electrification, and in a very sober tone. Although some articles did connect lightning threats to technological poles and wires, more common were articles about lightning and storms in general: experiences being struck, ways to avoid them (feather beds, suggested one article), and the havoc they can cause.[23] Others described workplace experiences of electrical shocks. Such stories were common, but did not progress to an anti-electrical discourse, or even suggestions of one. Instead of hyperbolic warnings of electrical threats, one encountered polite, business-minded criticism, such as this ad from a competing power company:

> ANDERSON, S. C., is known as the Electric City, but her leading hotel, the Chiquola, is lighted by Acetylene gas and why? Because it is cheapest and best. The Anderson Acetylene Light Co. installed for them a two hundred light "Ideal Eiworth Generator" which is simple, safe and perfectly automatic.[24]

This was in 1911, and no mention is made of the scare tactics famously used in the current wars of the 1890s. Indeed, the reaction to the health concerns of the fabled "current wars" is referred to as "ho-hum."[25] One of the most negative accounts I could find of electricity was in the form of a simple, humorous poem from 1910, whose curmudgeonly tone is anything but dystopian.

> To an Electric Light
>
> Twinkle, twinkle, little arc
> with your very sickly spark,

> Up above my head you swing—
> You're a very ugly thing.
>
> When the shades of night are come
> And I'm working hard at home,
> Then it is your "system" works,
> With its sputters, jumps and jerks.
>
> I don't care what people say,
> You've a bilious greenish ray,
> And wherev'er they worship you
> Everything is black and blue.
>
> Very likely you surpass
> Candles, oil and even gas:
> Still, allow me to remark,
> I don't like you, little arc!
> —La Touche Hancock[26]

Such mild humor, such as joking about the glare of electric lights off a balding man's pate causing blindness, were most benign and banal.[27]

Similarly, there was no utopian breathlessness. While there was civic boosterism in promoting electrification as a sign of and route to progress, it did not lapse into the expected hyperbolic claims. While there were many positive accounts of electrification, they tended to be staid accounts of installing new lines or a gas-electric railway. The most extreme language I could find was this headline and subheading: "Gas-electric cars: May be put into use by southern railway: Recent experiments have been entirely satisfactory and may mean practically a revolution."[28] Even the greatest source of potential hyperbole—advertising from electrical power companies—talked about savings and convenience in quiet, rational terms.

The longest story on electrical benefits I found was a reprint from the *New York Sun* in 1897. Running two full-page columns, it was headlined "Electricity's New Uses. Minor Needs of the Household Now Served By It." The article described electricity as advantageous and becoming fairly ubiquitous. However, rather than broad proclamations of curing diseases or liberating labor, the article listed several specific and quotidian new uses for electricity, in this order: door bell, hair curling, small stove for beverages or as a chafing dish, sewing machine, iron, foot warmer, heating pad, car, cigarette lighter, push buttons, intercoms, fans, hat cleaning, toy railway, and battery. Outside the home, the article listed a soldering iron, as well as specialized factory tools, such as drills and polishing devices, and looms, spindles, and carding machines in textile industries. The most upbeat language appeared in its conclusion:

> Everywhere that one sees the electric motor applied it becomes apparent that no other motor could compete with it for economy of space and the case with which power can be carried to it, and there are many situations where these qualities make it possible to be used when all other powers would be out of the question. With all of these advantages it has one other, which is of vital importance. This is the economy in the use of power. A good motor will return in actual work more than 95 per cent of the energy that reaches it, while no engine driven by other powers will approach this figure.[29]

No ethereal magic here, just a sound return on investment.

Accommodation, Not Reaction

Reactions to electrification and its attendant technologies are sparse in The Electric City. However, abandoning a perspective of linear causality that looks for things causing effects allows one to instead examine the various entanglements of ongoing, repeating processes of phenomena becoming. In Karen Barad's language of agential realism, this would exemplify intra-action, her replacement for interaction between separate things with the engagement of phenomena already connected and existing within the same universe. Although not as precise, a simpler way to think of this process is accommodation. Accommodation involves taking account for differences within a system, but they are already within that system. It is not inherently about a reaction to outside forces. Nor is it necessarily a one-time affair; it is ongoing.

For example, consider disability: There were always persons with atypical bodies and abilities in the United States. The Americans With Disabilities Act mandated that U.S. culture—technologies, architecture, transportation, commerce, etc.—accommodate them. While this changed both U.S. culture and the experience of being a disabled American, these were never separate things. They were part of the same system. Moreover, it was not a one-time event, but is a continuing process today. Both disabled Americans and American culture are continually entangled in the processes of each other's (re)becomings, through acts of accommodation. Also, rather than a paternalistic unidirectional act, I would suggest the two have always been in mutual entanglement and accommodation; the act of legislation just brings a moment of these processes into clarity.

Electrical forces were not new. What was new was their concrete status as thing, as commodity. Part of the process of ether becoming the utility bill involved accommodating this, integrating electrical current into the processes of daily life. Of this, The Electric City provides numerous creative and pragmatic examples, bereft of sublime over-reactions.

While textile mills were the most notable use of hydroelectric power, its application to other industries was a common topic, especially those related to the agricultural nature of the region, such as pest control.[30] An 1890 *Intelligencer* article reported on "beneficent results" of experiments in Colorado using electricity to slaughter livestock—a far cry from the spectacular displays of Thomas Edison's teams electrocuting an elephant and other animals as propaganda in the current wars.[31] In 1895, the *Intelligencer* reported on a case in nearby Greenville, SC,[32] in which electric lights "seemed to blind and confuse" a flock of wild ducks, leaving many "crippled and exhausted." Rather decrying technological disruption of the natural order, the article commends "enterprising citizens who went out early with guns [and] secured Sunday dinners without trouble."[33] The *Advocate* reprinted a *New York Times* article stating that not only crops, but also farmers needed "electrifying." "The growth of vegetation can be hastened by the use of electrical currents properly applied," it stated, and urged farmers to develop ways to do so with greater cost efficiency. Farmers needed "electrifying" because "as a class" they were behind the "state of the art" and knew their own business very poorly. Farm productivity would be increased if they could be appropriately "stimulated" and made aware of the opportunity before them.[34]

Other industries and practices were explored, but without breathless language of miracles or inherent magical properties of electricity. An article on electrical anesthesia in dentistry, while calling it a "remarkable success," described the public demonstration as merely "interesting." Electricity was not proclaimed a panacea, but clearly limited as something with "possibilities" in the dental field. It also described the use of an electric dental drill and mallet that "accommodates itself readily to the hand of the operator."[35] Other industries discussed included railroads and trolleys, office machines, electrical parts manufacturing telephony, and telegraphy. In an example of the latter, a 1910 article reported on a Harvard experimenter who used his iron bedframe instead of an aerial as an electrical receiver to receive wireless telegraph messages.[36]

(Re)Becomings

Bodies, minds, and spirits were already entangled with various electrical phenomena before industrial, commercialized electrification. This process merely separated and concretized existing phenomena in new ways. These were not encounters between new things, but shifting entanglements and accommodations of phenomena in ongoing, iterative processes of becoming. What, then, was becoming? I conclude this chapter suggesting how electrification was entangled with processes of forming a New South, racialization, gendering, and maintaining—not disrupting—existing power hierarchies.

New South

On the occasion of The Electric City's centennial, a local historian wrote, "In 1888, the city of Anderson caught the spirit of the new South, and begun [sic] to think and dream of whirling machinery, of busy communities where Anderson cotton could be manufactured into yards of cloth."[37] This "New South" was a vision of reclaiming former economic and cultural power and shedding the shame and scars of slavery, the Civil War, and Reconstruction. Technology was key to this process. Anderson and other areas of electrification represented an enthusiasm for technological modernization that could hopefully retain cultural values of the Old South while creating new prosperity to empower a New South.[38]

In 1893, Westinghouse had received the contract to illuminate the Chicago World's Fair, inspiring Andersonians to see electricity as key to progress.[39] One of Whitner's early employees was native Andersonian William States Lee, who worked as chief engineer on Portman Shoals. When Lee later became head of Duke Power Company and president of the Piedmont and Northern Railway, he recalled that his focus on hydroelectric power at the time had been because he knew that South Carolina had no coal, and without a native power source would remain primarily an agricultural state and would not progress into manufacturing.[40]

Press accounts at the time support this as well, and were very explicitly devoted to notions of progress. A visitor to a local paper in 1910 described it as unique for not being local, but aimed at presenting Anderson, its citizens, and "their progress" outward to the other citizens of the state.[41] Writers touted electrical Anderson over less advanced towns in the state: "Toni Pearce of Spartanburg is a live wire, and for that reason has almost decided to become a resident of the Electric City, although he has been doing his best to electrify the City of Distress for several years."[42] An 1819 article on "The Farmer of the Future" combined electrical implementations with elevation of the farmer himself in subsequent decades, with "far less chewing tobacco and spitting in the presence of others … [and] less profane language used."[43] The 304-page commemorative anniversary issue of *Anderson's Daily Mail*, from 1949, is titled "50 Years of Progress," and lists "epochal discoveries" demonstrating such progress in the region involving transportation, communication, engineering, and medical technologies, such as a "ten-minute brain operation, [the] transorbital lobotomy, combined with electro-shock was reported successful in treating some cases of schiziphrenia [sic] and involuted nail depressions."[44]

The South did change, but it was not the New South originally imagined. As Matthew Henderson has shown, the potential incompatibility between old values and new technologies were not immediately apparent. For example, electricity created a boom in the textile mills, but this, with electric illumination, helped create the structure of mill villages, small company towns, which ended up calling attention to class disparities, child labor, and other problems. Similarly the introduction of electric trolleys necessitated segregation, making racial stratification more visible, threatening the decorum of traditional antebellum reserve and gentility.[45] The South was becoming, but not the New South of which Andersonians may have dreamed.

Race

An oral history with a Jewish resident of Anderson presents a stark contrast to the press accounts of the sparkling "Electric City" at the turn of the century.

> Anderson, during that time, was a very hard town. It was mean, nasty, completely controlled by the Klan. The Klans, they kept the blacks in their place, and the Jews, they were tolerated, long as they stayed out of the way. We were second on the list. The blacks had a hard time because the poor white people—Anderson's a mill town and the mills were horrible. The kids worked in the mill when they were eight, nine, ten years old. The lint was so thick at the mill—I'd go down there—you could take your hand and just move it. People never lived past twenty-nine or thirty years old in those days. But the big mill owners, they kept the people in line; they kept the blacks in line and the poor whites in line, and the Klans helped them. The Klan was on the police department, they were in everything. Everybody knew who the Klan was. It was a sad thing.[46]

The maintenance of white superiority was not only used by used by electrification, technological progress was spoken of in similar terms. In rhetoric and actual actions—material-discursive practices—racism was part of the becoming of Anderson.

The Portman Shoals hydroelectric plant was located on the Seneca River, 11 miles west of Anderson, SC. The man-made Lake Hartwell, which replaced it, is now a recreation and retirement area for residents of the town of Seneca and nearby Clemson, home of Clemson University. "Seneca" derives from "Esseneca," which had been the name of a nearby, vibrant Native American trading community. The preeminent Anderson historian, whose work on electrification was the basis for a documentary on the History Channel, writes, "From an obscure town in the South Carolina Upcountry, the last area of the State to be settled after the Cherokee Indians gave up their homeland during the Revolution, it was transformed into a model city."[47] This transforming, this becoming, happened through electrification, but the region itself was a transformation of the original Cherokee lands. This was not a peaceful "giving up." European-American militias destroyed 50 Cherokee towns in 1776, and the remaining town leaders signed state treaties the following year.

The 40th anniversary commemorative issue of the *Daily Mail* writes that black and white laborers worked harmoniously together to build Portman Shoals, it ran in "absolute safety and perfect satisfaction," and "brought a new economic and social era to Anderson and the rest of South Carolina." Harmoniously does not mean equally, but in this case without conflict or incidence: stable, unthreatened white superiority. As electrification supported the spread of other technologies, the Advocate used racial stereotypes and medical misinformation to describe how Native Americans made ideal telegraphers. An article described how their "keenly developed sense of hearing" made them ideal telegraph operators. Furthermore,

> It has been found difficult to keep men at some of the lonely stations in the desert and mountain regions. Indians would not object to being alone in these places. … Superintendent says they enjoy it and do good work. "They ought to be especially successful because of their repose and lack of nervousness under the strain of work." This pressure drives many white men out of the business prematurely because their health has given way.[48]

Recalling the earlier discussion of neurasthenia, here again we see medical pathologization used to support racial hierarchies: The less sensitive, civilized Native American operators were not susceptible to emotional and nervous disorders, as were the "savage" races discussed by George Miller Beard.

The early period of electrification overlapped with the removal of African-American congressional representatives and senators elected during Reconstruction. Between 1890 and 1908, Southern state legislatures passed laws and constitutional amendments with voting restrictions that served to disenfranchise African-American voters. Violence and terrorism, as noted earlier, were also methods of disenfranchisement and social control. In this sense, the South was socially attempting to re-become an even Older, pre-Reconstruction South. A suggestion of this entanglement of old and new appears in the description of electricity and its technologies as, not outside threats or saviors, but humble "servants."[49]

In 1975, Anderson's centennial, a local historian published a collection of significant episodes from the town's past, such as Portman Shoals. As such, it continued the linkage between technological and scientific accomplishment and social progress, except, this time, in a retrospective way, describing past accomplishments. The civil rights movement had forced change on the local culture, with nearby Clemson University reluctantly accepting its first African-American student in 1963. Moreover, on top of national economic crises, during this period South Carolina began to experience the decline of agricultural and textile industries. Within this context, one of the claims of progress Anderson made was the discovery of anesthetic ether.

Dr. Phillip A. Wilhite, a Georgia native, was a leading doctor in Anderson for 40 years and frequent vice president of South Carolina Medical Association. (He declined to be president.) He died in 1892. As related by his close friend, Dr. J. Marion Simms of New York City, he was an "unsung hero" for his discovery of the use of ether as anesthetic. The discovery is described as part of a "quilting." Quiltings were common entertainment in the early 1800s, and young men and boys would go hoping for a dance or "ether party."[50] In the fall of 1838, Wilhite was 17, and inhaling ether after a quilting with other boys and girls. He saw an African-American boy watching them at the door, presumably a slave, and invited him in to try ether. He refused, but they insisted multiple times. Sims relates:

> He struggled violently, but they threw him down and held him there while Wilhite poured out some ether on a handkerchief and pressed it firmly over his mouth and nose. He fought furiously. They persisted, thinking it was great fun.
>
> After a long struggle, the boy became quiet and unresisting. … He lay quietly and with stentorous breathing. They tried to arose him, but could not.

> They then became greatly alarmed, and sent one of their number on horseback for Dr. Sydney Reese at Athens.

By the time the doctor arrived, Wilhite and his "principal accomplice" thought they had murdered the boy and were considering escaping the country. They told the doctor what happened, and he slapped and shook the boy until awake. In the years soon after, Wilhite went to medical school, graduated in 1852, and began a practice. At a meeting of the American Medical Association, Wilhite was recognized as having first uncovered these effects of ether. A committee was formed to petition Congress to provide him with a pension for his service. Dr. Sims went on to give Wilhite full credit in an issue of *The Druggists' Circular*.

Here we see how racial power was used to generate scientific knowledge, using an African-American as an experimental subject in a pattern familiar from Henrietta Lacks, the Tuskegee Experiment, and "father of gynecology" J. Marion Sims. However, in 1975 it was cited retrospectively and ceremoniously as a claim for not what Anderson could be, but what it was and had been.

Gendering

Gendering appears in the discourses of electrification as well. Reporting on the increased danger of lightning in cities with electric wires, the *Intelligencer* demonstrated entangled pathologization and feminization, arguing that, "The actual danger [from an electrical storm] is, in truth, not from the lightning nor the thunder, but from the nervous condition into which women allow themselves to fall." Advertisements for telephone service reinforced pathological feminine stereotypes by promising to solve the problems of lonesome and vulnerable female telephone users.[51] As discussed previously, neurasthenia was a nervous condition brought on by modern technologies, women's education, and other aspects of modernity. A 1910 *Advocate* article described a similar pathology, but in this case modern changes, such as the women's movement into higher education and the workplace (home of electrical office machines, telegraphs and telephones) was causing "civilized" women to physically evolve into more mannish, masculine bodies.[52]

Moreover, a science of female biology—gynecology—played a significant role in the electrification of Anderson. The "Father of Gynecology," Dr. J. Marion Sims, was from Lancaster, SC,—and had related the previous historical episode regarding ether. Sims also had convinced Walker Gill Wylie to switch from engineering to medicine, specializing in gynecology. Wylie was also a South Carolinian, and had studied civil engineering as an undergraduate at the University of South Carolina. He had moved to New York City for medical school shortly after the Civil War, and remained there for the rest of his life, becoming one of the leading surgeons in the nation and the first doctor to train women as operating room nurses, actively encouraging and supporting their employment as a corrective to the neglect and brutality of male nurses at the time, much as female telephone operators had replaced unruly boys. Wylie attended a lecture by William Whitner on public health (which, at the time, often included hygiene, sexually transmitted diseases, and other areas related to gender), which led to a partnership on developing mills in South Carolina. When Whitner was experimenting and developing Portman Shoals, one of his strongest local supporters had been Dr. Samuel Orr, who specialized in "female complaints" and was popular and sought-after throughout the state. His success was attributed to his great "personal magnetism" that put clients and others at ease. Recall that, as discussed earlier, concepts such as "personal magnetism" initially had direct, physical, electrical conceptions, not merely the idea of personality or charisma we would think of today. The science and scientists of feminine pathology were key to Anderson becoming The Electric City.

Power Hierarchies

Cause-and-effect reactions, and the object-oriented ontologies they support, emphasize a sense of change, making it difficult to see processes of maintenance. In contrast, continual processes of becoming may involve change—or not.

Born in Anderson, Samuel Orr was the son of an ex-governor of the state and brother of a state legislator, James, who had his own mill business in the 1890s and was a national leader in the industry. The Orr brothers were powerful men from a powerful family. Several sources assert that they, especially Samuel, were a major influence in convincing people to support Whitner and maintaining public confidence.[53] As one writer said, "They wrought well a greater state."[54] After retiring from practice, Samuel became President and Treasurer of the Board of Directors of Anderson Water Power and Light Company, which owned Portman Shoals. Power maintains power.

During this time period, electricity was also involved with the most basic exercise of power: killing. During the current wars, public safety was the increasing theme of DC supporters, and they held public displays using AC current to injure and kill animals. Thomas Edison's DC supporters championed the use of AC in the electric chair as a new form of capital punishment.[55] In 1890 the first death sentence by electrocution was handed out in Auburn, New York, which Westinghouse protested. It received wide publicity but, "As for the public, which was being asked by the direct current forces to judge the proceedings, the issue was not nearly as serious as might have been supposed. The execution was carried out and the public returned a verdict that could be interpreted as 'Hohum,' to say the most. The emotions had been drained from Edison's argument."[56] Executions using AC continued around the country to an "indifferent" public, but so did expansions of AC systems, such as the controversial award of the Chicago World's Fair contract in 1893 to Westinghouse.[57]

Anderson had a nearby equivalent. As reported in the *Advocate* in August of 1910, a judge in Asheville, North Carolina, sentenced murderer James B. Allison to death by electrocution before a packed courtroom, just months after the execution in Auburn. While in prison, Allison unsuccessfully attempted to cut his own throat to avoid death by electrocution, but was executed in 1911.[58] Electrification brought a new iteration of the state becoming an entity with power over life and death, but not a change in that power relation.

The wealth of the South was agricultural, founded and dependent on power relationship of slavery. An economic engine that supported the nation, primary crops included cotton and tobacco. Tobacco farming made the Duke family rich and powerful, and another battle of power—the current wars between Edison and Westinghouse, grabbed their attention. In 1898, they began buying up hydroelectric sites in the Carolinas. In 1904, Buck Duke was stuck in bed with foot problems, just as Whitner had been recuperating in bed with illness when Anderson approached him about his first project. This time, Duke was the captive audience to Dr. Gill Wylie. Wylie had left his brother to manage their hydroelectric projects in South Carolina, developed with Whitner, and returned to medical practice in New York. Wylie told Duke about his experiences with hydroelectric power with Whitner, furthering his interest. The two men went on to partner with William States Lee to form the first network of hydroelectric plants that would become what is today Duke Energy, the largest electrical power holding company in the United States.

After 20 years, when it was time for the Anderson City Council to renew the electric utility franchise, and they did so without a public vote. They kept the existing power structure in place and justified this by the fact that Anderson had become The Electric City:

When the old franchise was granted Anderson was a village, or small town. A few kerosene lamps scattered here and there lighted it. The franchise was granted and Anderson began to grow, becoming known as "The Electric City." The abundance of current and advantages offered induced many manufacturing plants to come here, and many thousands were added to her population. Therefore when the time for renewing this franchise came, twenty years later, there were different conditions surrounding the city and some changes had to be made. But, we venture the assertion that the rights of the city are better guarded under the present franchise than they were under the old.

But, our contention is that council had a right to grant a "renewal" franchise, as it did a few months ago, without holding an election to validate it.[59]

Postscript: Becoming

The statue of Whitner and the Confederate memorial stand almost side-by-side on Main Street in downtown Anderson (Fig. 4.3). Downtowns are

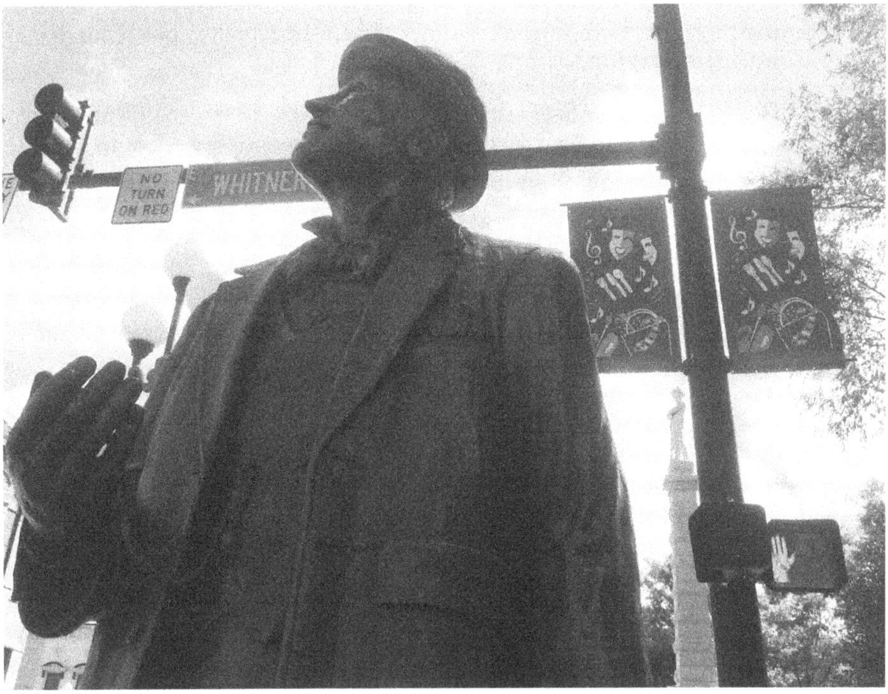

Figure 4.3. Complex history of the old/new South. Statue of electrical innovator William C. Whitner and street sign named after him in foreground. Memorial to Confederate soldiers in background. Main Street, downtown Anderson, SC, 2014.
Source: Photo by author.

important in the Upstate, with nearby Greenville—my home—receiving international attention for its downtown revitalization, a key to economic development through tourism and outside investment that has slowly built a recovery from the textile industries that were largely outsourced during late twentieth-century neoliberal globalization of industries. Historic downtowns are rebecoming central to places like Anderson, although Upstate businesses continue to face problems with recruiting and retaining outside talent, due to perceptions of racism, sexism, and generally backwards culture.

I certainly considered these things when, as one-half of a gay male couple legally married in California in 2008, I was offered a job at Clemson University, 18 miles north of Anderson. In fact, upon accepting the job, we chose to live in Greenville, 30 miles east of the University, as it seemed to be a larger, more urban, and less potentially homophobic place than the small town of Clemson. In working with queer students and issues here, Clemson has become more progressive, and I have become less prejudicial in my thinking about rural communities. The ongoing processes of our mutual becomings have entangled.

Clemson, despite being a land-grant, public university, has a far lower minority enrollment than the area population and even all other colleges and universities in the state. Recently, the university has been reeling from faculty who were threatened and ostracized in various ways for engaging in the protests over the shooting of an unarmed black youth in Ferguson, Missouri, and on the recent "Black Friday" after Thanksgiving. Against this backdrop, an all-white Clemson fraternity threw a "Cripmas" party at which participants dressed in black gangsta costumes. Meanwhile, an African-American graduate student, AD Parsons, started See The Stripes, a program of racial dialogue and visibility, and the university has organized an ongoing series of events on "Race and the University." I played a small role in the organization of that series, and, inspired by its events, took my class of graduate students in technology studies to see the oldest building on campus, whose foundation is made of stones from the previous plantation's slave quarters, and built by convict labor, many of whom were African-American juveniles. After a student protest, Clemson has made progress in acknowledging such history.

Since being here, I've experienced insights into the entanglements of my own privilege and power as a white man. My aunt recently uncovered that we had ancestors who had lived in Boone, North Carolina, a mountain town three hours north of where I live. They were a successful, powerful, merchant family, building the general store that founded the town and marrying into the Mast family, of Mast General, a regional chain of high-end stores for tourists.

My aunt also found online digitized copies of census documents listing our relatives' slave holdings. They were a large family, and we still don't know how or why our branch of it went to Texas, where we are from, but it is safe to assume that their power and resources enabled them to do so. There is a street bearing my mother's maiden name in downtown Boone.

That family name, Council, is also shared with the judge who issued the first death sentence by electrocution in this region, which I discovered while researching this book which, when published, will contribute to my becoming a full professor at Clemson. This will help me become more secure not only financially but in response to the challenges of having a severely mentally ill mother, one who exhibits the many pathologized feminine attributes and stereotypes described throughout this book. Indeed, my life experiences as a pathologized gay man coming of age during the AIDS epidemic influence my interests in this book as well.

I conclude this chapter on such a reflexive note to illustrate complexity. Social forces do not work in clean, tidy, linear relationships of cause and effect. Individuals, audiences, users, diseases, technologies, social groups, genders, and races: none of these are things; they are phenomena constantly in ongoing processes of becoming and rebecoming. We are action. We are motion. And moving targets are complicated.

This book thus far has focused on mapping out the complicated, moving targets of pathological technoculture. In subsequent chapters, I turn mostly to specific case studies of technopathologies, demonstrating their staying power and cultural work.

Notes

1. Nye, *American Technological Sublime*, 390.
2. Foucault, "Infamous Men," 172.
3. Carey, *Communication*; Lanza, *Elevator*, 33; Marvin, *When Old Technologies*; Marx, *Machine*; Metzger, *Blood and Volts*; Milutis, *Ether*; Mosco, *Digital Sublime*; Nye, *American Technological Sublime*; Sconce, *Haunted Media*; Simon, *Dark Light*; Winner, *Whale*.
4. Simon, *Dark Light*.
5. Biggs, "Pleasure Telephone," 61.
6. Simon, *Dark Light*, 72.
7. Marvin, *When Old Technologies*, 233.
8. Brox, *Brilliant*; Nye, *Electrifying*; Schiffer, *Power Struggles*.
9. *40th Anniversary*; "City's Growth;" "Electric City Traces Transmission Routes;" "Electric Epigrams;" Klosky, *Six Miles*; "Portman Dam and Power Plant;" "Power from Portman

Shoals;" "President Will Pay a Visit to Anderson;" Dickson, *Journeys*; Watkins, *Anderson County*.

10. Tabler, "South Carolina."
11. "The Electric City," 1.
12. Griffith, "Man for all Seasons."
13. Brox, *Brilliant*.
14. "A Curious Prophecy," 2.
15. "Insomnia," 4.
16. *40th Anniversay*, also quoted in Klosky, *Six Miles*, 89.
17. "Darlington Herald," 1.
18. "Electric Bitters," 1; "Enpepsy," 1.
19. "Electroplating Human Bodies," 2; "Susceptibility to Disease," 4; "Mind Cures," 4.
20. "Bad Moving Pictures …," 1; "Would Kill the Mayor …," 5.
21. "Street Lighting Novelty," 3.
22. "Look Out for the Electric Storms," 6.
23. "Avoiding Electrical Accident …;" lightning's queer pranks, 2; "Insomnia," 4.
24. "Anderson S. C. is known …," 8.
25. Maynor, *Duke*.
26. Hancock, "To an Electric Light," 7.
27. "The Chicago Theatre-Going Ladies …," 1.
28. "Gas-Electric Cars," 2. See also "A New Line of Wires."
29. "Electricity's New Uses," 1.
30. "New Way to Kill Bugs," 8.
31. "One of the Beneficent …," 1.
32. The town where I have resided since 2010.
33. "There was an Enormous …," 1.
34. "To Electrify the Farmers," 8.
35. "Electricity in Dentistry," 1.
36. "Wireless in His Little Bed," 3.
37. Van Diver, *Journeys*, 293.
38. Henderson, "Reflections".
39. Klosky, *Six Miles*, 54.
40. Dickson, *Journeys*.
41. "Editor Grist," 3.
42. "In My Town," 6.
43. "The Farmer of the Future," 4.
44. 50 *Years*, no page number.
45. Henderson, "Reflections".
46. Siegel, "Oral History," 2.
47. Klosky, *Six Miles*, 67.
48. "Indian Telegraphers," 7.
49. *40th Anniversary*; Dickson, *Journeys*, 156; Stern and Gwathmey, *Once Upon*.
50. Dickson, *Journeys*, 101.
51. "To Lonesome Women;" "A Constant Protection."

52. "Woman Like Man," 2.
53. "They Wrought;" Klosky, *Six Miles*; Van Diver, *Traditions*.
54. "They Wrought," page not numbered.
55. Metzger, *Blood and Volts*; Essig, *Edison*.
56. Maynor, *Duke*, 10.
57. Klosky, *Six Miles*, 39.
58. "Will Be Electrocuted," 1; "Tried to Cut Throat," 1.
59. "A Renewal Franchise," ¶1–2.

References

40th Anniversary and World's Fair Edition. The Anderson Daily Mail. Anderson, SC. Commemorative edition, 1939, April 29.

50 Years of Progress 1899–1949. The Anderson Daily Mail. Anderson, SC. 304 unnumbered pages. Commemorative edition, 1949, November 5.

"A Renewal Franchise." *The Anderson Daily Intelligencer*. 1914, November 12. Library of Congress Chronicling America database.

"Anderson S. C. is known. …" Anderson Acetylene Light Co. advertisement. Newberry S. C.: *The Herald and News* 1911, February 21, p. 8. Library of Congress Chronicling America database.

"Avoiding Electrical Accident. …" *Anderson Daily Intelligencer*. Library of Congress Chronicling America database.

"Bad moving pictures. …" *Anderson Peoples Advocate*, 1910, July 11, p. 1.

Biggs, Asa. "The Pleasure Telephone: A Chapter in the Prehistory of Media." In *The Social Impact of the Telephone*, edited by Ithiel de Sola Pool, The MIT Press, 1977.

Brox, Jane. *Brilliant: The Evolution of Artificial Light*, Houghton Mifflin, 2010.

Carey, James W. *Communication as Culture: Essays on Media and Society*, Unwin Hyman, 1989.

"The Chicago Theatre-Going Ladies. …" *Anderson Daily Intelligencer*, 1897, April 7, p. 1.

"City's Growth." *40th Anniversary and World's Fair Edition. The Anderson Daily Mail*. 1939, April 29, no page number.

"A Constant Protection." *Anderson Peoples Advocate*, vol. 20, no. 101, 1910, October 27, p. 8.

"A Curious Prophecy." *Anderson Peoples Advocate*, 1890, November 17, p. 2.

"Darlington Herald." *Anderson Daily Intelligencer*, 1893, April 18, p. 1. Library of Congress Chronicling America database.

Dickson, Frank. *Journeys into the Past: The Anderson Region's Heritage*. Anderson, SC: Anderson County Bicentennial Committee, 1975.

"Editor Grist of Yorkville Writes up Trip to Anderson." *Anderson Peoples Advocate*, 1910, August 8.

"Electric Bitters." *Anderson Peoples Advocate*, vol. 20, no. 48, 1910, April 21, p. 1.

"The Electric City." *Anderson Daily Intelligencer*, 1895, June 19, p. 1. Library of Congress Chronicling America database.

"Electric City Traces Transmission Routes." *40th Anniversary and World's Fair Edition. The Anderson Daily Mail*, 1939, April 29, no page number.

"Electricity in Dentistry." *Anderson Daily Intelligencer*, no date, circa 1890s, p. 1.
"Electricity's New Uses." *Anderson Daily Intelligencer*, 1897, July 14, p. 1.
"Electroplating Human Bodies." *Anderson Peoples Advocate*, 1890, December 22, p. 2.
"Enpepsy." *Anderson Daily Intelligencer*, 1891, February 12, p. 1. Library of Congress Chronicling America database.
Essig, Mark. *Edison and the Electric Chair: A Story of Light and Death*. Bloomsbury Publishing USA, 2005.
Foucault, Michel. "Lives of Infamous Men." In *The Essential Works of Michel Foucault: Vol 3: Power*, edited by Paul Rabinow, Penguin, 2002, pp. 1954–1984.
"Gas-Electric Cars." *Anderson Peoples Advocate*, vol. 20, no. 60, 1910, June 2, p. 2.
Griffith, Steve. 2008. "A Man for All Seasons: Walker Gill Wylie." *Caroliniana Columns*, no. 23, pp. 8–12.
Hancock, La Touche. "To an Electric Light." *Anderson Peoples Advocate*, 1910, July 27, p. 7.
Henderson, Matthew. "Reflections on Electricity, Modernization, & Identity in the New South." Master's Thesis, Clemson University, 2011.
"In My Town," *The Anderson Daily Intelligencer*, 1914, June 30, p. 6. Library of Congress Chronicling America database.
"Indian Telegraphers: Have Keen Hearing and Touch and Do Not Get Lonely." *Anderson Peoples Advocate*, 1910, November 28, p. 7.
"Insomnia." *Anderson Peoples Advocate*, 1891, August 17, p. 4.
Klosky, Beth Ann. *Six Miles that Changed the Course of the South, the Story of the Electric City, Anderson, South Carolina*. The Electric City Centennial Committee, 1995.
Lanza, Joseph. *Elevator Music: A Surreal History of Muzak®, Easy-listening and Other Moodsong,®* revised and expanded edition, University of Michigan Press, 2004.
"Look Out for the Electric Storms to be Very Bad This Summer." *The Manning Times*, 1913, June 25, p. 6. Library of Congress Chronicling America database.
Marvin, Carolyn. *When Old Technologies were New: Thinking about Electric Communication in the Late Nineteenth Century*, Oxford University Press, 1988.
Marx, Leo. *The Machine in the Garden: Technology and the Pastoral Ideal in America*, Oxford University Press, 1976.
Maynor, Joe. *Duke Power: The First 75 Years*, Delmar Press/Duke Power, 1979.
Metzger, Tom H. *Blood and Volts: Edison, Tesla, and the Electric Chair*, Autonomedia, 1996.
"Mind Cures." *Anderson Peoples Advocate*, 1891, February 2, p. 4.
Milutis, Joe. *Ether: The Nothing that Connects Everything*, University of Minnesota Press, 2006.
Mosco, Vincent. *The Digital Sublime: Myth, Power, and Cyberspace*, The MIT Press, 2004.
"New Way to Kill Bugs." *Anderson Daily Intelligencer*, 1914, July 15, p. 8. Library of Congress Chronicling America database.
Nye, David E. *American Technological Sublime*, The MIT Press, 1994.
Nye, David E. *Electrifying America Social Meanings of a New Technology, 1880–1940*, The MIT Press, 1992.
"One of the Beneficent. …" *Anderson Daily Intelligencer*, vol. 25, no. 9, 1890, September 4, p. 1.
"Portman Dam and Power Plant." Historical marker in Generator Park, Anderson, SC. Corner of McDuffie and Whitner Streets.

"Power from Portman Shoals." *Quills Employee Magazine for Abney Mills*, vol. 26, no. 11, 1977, September, p. 2.

"President Will Pay a Visit to Anderson." *Anderson Daily Intelligencer*, 1911, February 3, p. 3. Library of Congress Chronicling America database.

Schiffer, Michael Brian. *Power Struggles: Scientific Authority and the Creation of Practical Electricity Before Edison*, The MIT Press, 2008.

Sconce, Jeffrey. *Haunted Media: Electronic Presence from Telegraphy to Television*, Duke University Press, 2000.

Siegel, Sam. *Oral History Interview*. Mss. 1035–062 Jewish Heritage Collection, College of Charleston, 1996.

Simon, Linda. *Dark Light: Electricity and Anxiety from the Telegraph to the X-ray*, Harcourt, 2004.

Stern, Ellen, and Emily Gwathmey. *Once Upon a Telephone: An Illustrated Social History*, Harcourt Brace and Company, 1994.

"Street Lighting Novelty." *Anderson Peoples Advocate*, 1910, September 19, p. 3.

"Susceptibility to Disease." *Anderson Peoples Advocate*, 1890, December 29, p. 4.

Tabler, Dave. 2012. "The South Carolina Man Who Put the Electric in 'The Electric City.'" *Appalachian History*. August 24. http://www.appalachianhistory.net/2012/08/south-carolina-man-who-put-electric-in.html. Accessed July 12, 2015.

"There was an Enormous. …" *Anderson Daily Intelligencer*, 1895, May 1, p. 1.

"They Wrought Well a Better State." *The Anderson Daily Mail*, 40th Anniversary and World's Fair Edition, 1939, April 29, no page number.

"To Electrify the Farmers." *Anderson Peoples Advocate*, 1910, August 22, p. 8.

"To Lonesome Women." *Anderson Peoples Advocate*, vol. 20, no. 93, 1910, September 29, p. 3.

"Tried to Cut Throat." *Anderson Peoples Advocate*, 1910, August 25, p. 1.

Van Diver, Louise Ayer. *Traditions and History of Anderson County*, Anderson Daily Independent, 1928.

Watkins, William Law. *Anderson Country, South Carolina: The Things that Made it Happen*, Anderson County Museum, 1995.

"Will be Electrocuted: Asheville Murderer Sentenced to Death by Judge Council." *Anderson Peoples Advocate*, 1910, August 11, p. 1.

Winner, Langdon. *The Whale and the Reactor: A Search for Limits in an Age of High Technology*. Chicago: University of Chicago Press, 1986.

"Wireless in His Little Bed." *Anderson Peoples Advocate*, 1910, October 20, p. 3.

"Woman Like Man." *Anderson Peoples Advocate*, 1910, December 1, p. 2.

"Would Kill the Mayor. …" *Anderson Peoples Advocate*, 1910, July 7, p. 5.

· 5 ·

NOT SO NEW

Historic Continuity and the Pathologization of Users

> In the tangled networks of a great city, the telephone is the unseen link between a million lives … It is the servant of our common needs—the confidante of our inmost secrets … life and happiness wait upon its ring … and horror … and loneliness … and death!!
>
> —Opening captions of *Sorry, Wrong Number*, 1948

> TV causes Eye Problems in 40 percent of American Preschool Children, Optometrist Claims
>
> —1972 tabloid headline

In 2013, the campus magazine of my university ran a cover story on "Techno-stress," co-authored by "Varun Grover … the William S. Lee Distinguished Professor in Information Systems at Clemson. He ranks eighth out of 400 prominent management information systems researchers in the 2012 update of the University of Arizona's h-index ranking, widely accepted as the metric that assesses the productivity and impact of a scientist or scholar." The article began:

> Do you get concerned about losing your cellphone or mobile device?
>
> Are you anxious if there is no wireless service?

> Do you feel your life is dependent on Internet availability?
>
> If you answered "yes" to any of these questions, then you might be suffering from "nomophobia"—fear of having no mobile phone.
>
> Or, if you felt that your phone vibrated when it didn't, you might be experiencing "phantom vibration syndrome."
>
> Those are just two examples of the adverse side effects of our interactions with today's technologies. Unfortunately, these effects, as well as others, are becoming widespread as contemporary information technologies get increasingly embedded into our day-to-day lives.
>
> The psychological and physiological impact of these effects is broadly coming under the umbrella term of "technostress." High levels of technostress can lead not only to significant chronic health ailments, but also to decline in productivity levels.

Actually, 'technostress' was the title of a 1984 book, published 29 years earlier, on the human cost of the computer revolution.

In the previous chapters, I demonstrated the historic and cultural logics behind technopathologies, and provided a counterexample to the topian narratives they are often ascribed in the case of Anderson, SC. Examining themes across multiple examples, this chapter will expand my argument on two fronts. One, I continue the complication of newness begun in the previous chapter on Anderson by articulating and refuting the *newness hypothesis*: that technopathologies are reactions to new media and technologies that can be interpreted as reactions to other social changes. Through multiple examples, I will demonstrate that technopathologies have been associated with dominant and even old technologies. Two, I argue this indicates the *individualization effect* of technopathologies. Pathological technoculture is not about the demonization of scary, threatening technologies, but the pathologization of sick, unhealthy users. Moreover, we see how technopathologies shift responsibility away from social and systemic factors to the individual responsibility of users. The chapter concludes with an example of contemporary entanglement with a very old technology, electricity, in the technopathology of electrosensitivity.

The Newness Hypothesis

Newness has been overused as a theoretical construct. It provides an easy out for more complex analyses of technology. Although a rich body of scholarship has interrogated and complicated the concept of newness,[1] my project here is not to further destabilize or queer newness, but to unpack some of the thinking associated with it.

In researching, presenting, and discussing technopathologies, I have frequently encountered what I refer to as the newness hypothesis. This consists of two components. The first is the association of technopathologies largely or exclusively with new technologies. For example, a review of scholarship on "Internet addiction" links it to a previous half-century of concerns over new communication technologies, such as television and video games, and their impacts on social and individual well being.[2] A similar study focused on the Internet's sex-related concerns, such as addiction, psychological damage, abuse, rape and moral decay, and detailed similar fears of threats to physical, emotional, moral, social and sexual health due to sexual and erotic uses of print media, movies, popular music and music videos, telephones, and computers.[3] Both studies placed their phenomena as current or common reactions to new technologies that may be valid, but are not entirely new or different; it is more a question of quality or degree. However, the maladaptive behaviors are always in reaction to technological evolution and change; it is the *newness* that is the source of the problem.

The second aspect of the newness hypothesis is the *interpretation* of technopathologies as expressions of irrational fear of change as represented by new technologies. The diseased body or mind here is merely a conceptual tool used to translate and interpret the new and unfamiliar. In her history of the relationship between electricity and anxiety, Linda Simon describes fears related to electricity and one's health and well-being. She cites one writer's fear that electricity running through telegraph and telephone lines would cause "apoplexy, palsy, paralysis of the brain, and nervous disorders so profound that death would be a welcome remedy." Simon writes that

> Fear of electricity may seem quaint to us. … Yet, truth be told, we, too, worry about invisible emanations—from power lines or from the cell phones we carry with us everywhere. New technologies unsettle us. … Many of us have inherited from our nineteenth-century forebears some cherished assumptions: that nature will and must keep its secrets, that humans should not transgress, that nature will enact punishment against those who try to control it or probe too closely.[4]

As new technologies were developed, so were new maladies: a 1989 book-length study on computer addiction (or "dependency") contextualized it with fears of other mental illnesses, such as anxiety, stress, fear, social isolation and withdrawal, alienation, and phobias; as well as physical health threats, such as cataracts, facial rashes, eye strain and damage, epilepsy, miscarriages, birth defects, swollen joints, and poor posture. While admirably historicizing these technopathologies through acknowledging earlier fears of television addiction and telephone deafness, the study asserts that "computer dependency was only one of a multitude of computer-related anxieties … [and] found to differ little from the anxieties which surrounded the introduction of many other technologies over the centuries."[5] Yet, a full 22 years later, Sherry Turkle's *Alone Together: Why We Expect More from Technology and Less from Each Other*, sees the digital age provoking alienation and diminished direct human contact. Although rejecting the construct of addictions as being too disempowering, she nevertheless describes many familiar addictive and pathological symptoms such as compulsivity, antisocial withdrawal, anxiety, and narcissism as due to digital technology use. In 2014 a wearable internet device, Google Glass, was blamed for causing internet addiction.[6] In 2016, Andrew Sullivan, an early champion of political blogging and one of its first superstar pundits, wrote a lengthy confessional of his internet addiction and "distraction sickness" titled "I Used to be a Human Being." He describes how, during his final year of blogging, his physical health was so bad that his doctor asked Sullivan—one of the earliest and biggest stars of new media—"Did you really survive HIV to die of the web?"

This is nothing new. The addictiveness of computers remains still the central conceit, merely updated through iterations of applications, networking, and peripheral devices. A 2014 *New Yorker* article, "Is Internet Addiction a Real Thing?", noted that the addictive potential of the web dated back to at least 1997.[7] By the 2010s, wide-spread access to the Internet was approximately 20 years old, and interaction with personal computers easily twice that.

Often technopathologies appear as sidenotes within media histories and social analyses. Carolyn Marvin describes how the once-new force of electricity "could be an ominous symbol of projected anxieties, a signifier not of order but of instincts indulged at peril," with popular reports describing increased health risks from electrical shocks and accidents, new forms of auditory strain, tinnitus, hallucinations and nervous insanity from the telephone, eye inflammation from electric lights, and weakened nerves from the telegraph. These cases of electricity "as a transmitter of bodily maladies" are discussed as part of a wide range of fantastical reactions to electricity, including miracle healings, superhuman powers, intelligent machines, utopian and dystopian visions, and entertainment spectacles, an "enormous range of discussion about electricity, nature, and the body[, which] attempted to locate electricity, a force of unknown dimensions, by means of the most familiar of all human landmarks, the human body."[8] Lynn Spigel's *Make Room for TV* has references to concerns that, as television was adopted, it was part of a drug-like mass media that could contribute to overexcitement, addiction, damage to eyesight, moral corruption, facial tics, violence, and isolation, particularly for children. TV was also discussed using pathological metaphors, such as "telebugeye" and "spectatoritis." Yet, she asks in her introduction: "If television was believed to cause childhood maladies, might this not be symptomatic of the country's larger concerns about problem children and juvenile delinquents?"[9] Indeed, the technopathologies throughout her book are not examined together, but as varied examples of ways in which reactions to television reflected concerns about changing American ideals of family, gender, and social space in the postwar era.

While such analyses offer useful cultural insights, they limit our understanding of technopathologies primarily in terms of new technologies as symbols of social change. As I will illustrate, technopathologies also occur when new technologies have become commonplace, and when common technologies have become old. Technopathologies are not only about the new. And, as I described in Anderson, new technologies and social changes do not always incur technopathologies and topian reactions. Thus, technopathologies cannot only be symptoms, metaphors, visions, fears, or anxieties about new social configurations. While I do not disagree with the analyses of these and similar scholars, my point is that there is more to pathological technoculture than new media. The interpretive impetus of the newness hypothesis may provide many insights, but it also forecloses the possibility of taking technopathologies seriously.

Not So New

If technopathologies were merely expressions of anxious reaction to new technologies, they would go away as technologies were widely adopted. Yet, pathologies have been associated with electric communication technologies that are no longer new. Artificial illumination was associated health concerns such as over-eating in 1983.[10] To this day, concerns persist about light pollution and its impact on the health of humans and other animals.[11] The telegraph was the first global network of communication connections, what historian Tom Standage called "The Victorian Internet." Although the first transatlantic cable was active in 1858, a telegraphy injury, "glass arm" or "telegrapher's paralysis," was being discussed as late as 1976.[12] Radio, initially referred to as "wireless telegraphy" and often used for one-to-one and amateur, as well as mass-broadcast, communication, became widespread in the United States in the early 1920s. And yet, as late as 1965, *Pastoral Psychology* reported on a meeting of the Acoustical Society of America at which transistor radios were listed as a contributing factor to "urban sound."[13] This condition was seen to "contribute to circulatory troubles, loss of hearing, fatigue and emotional disturbances, nagging, and wife beating."[14] In 1983, neither telephony nor television were new, yet an article in the *National Enquirer* described how to properly use them to avoid headaches.[15] As late as 1991, a NYC doctor coined "Free Hand Syndrome" to describe an "ailment affecting the neurovascular bundle in the brachial-plexus region of people who talk on the telephone while simultaneously opening mail, cooking, taking notes, etc."[16] As late as 2008, a cable television program, *Stranger Danger*, devoted to debunking popular myths, still felt the need to refute the technopathology of television causing eye damage, especially in the young. The American Optometric Association website currently contains a webpage devoted to the causes, diagnosis, and treatment of "Computer Vision Syndrome," although computers have been widely used for decades.

This persistence extends to the common force in these technologies: electricity. Electricity is no longer new, so it should not make users sick. And yet, in 1985, an orthopedic surgeon at SUNY Upstate, working for the Veterans Administration, recounted his experiences as a member of an expert committee evaluating the physiological hazards of various electromagnetic pollutions. He presented research data in a widely reviewed book that "electromagnetic pollution" has been associated with depression and deleterious physiological effects worse than previously thought.[17] A 1994 General Accounting Office report to the House of Representatives' Committee on Natural Resources noted that, despite no firm scientific evidence, 11 states had taken steps to reduce public exposure to electromagnetic fields.[18]

One of the most widely known technopathologies is the association of health threats with cell phones and their towers (Fig. 5.1). A book-length study of the phenomenon linked it to a history of fears of new technologies, particularly those involving invisible forces such as radiation, microwaves, X-rays, and power lines, and attributed to social developments, such as Americans' "exaggerated reaction[s] to everyday risk," precaution-based culture, and media sensationalism.[19]

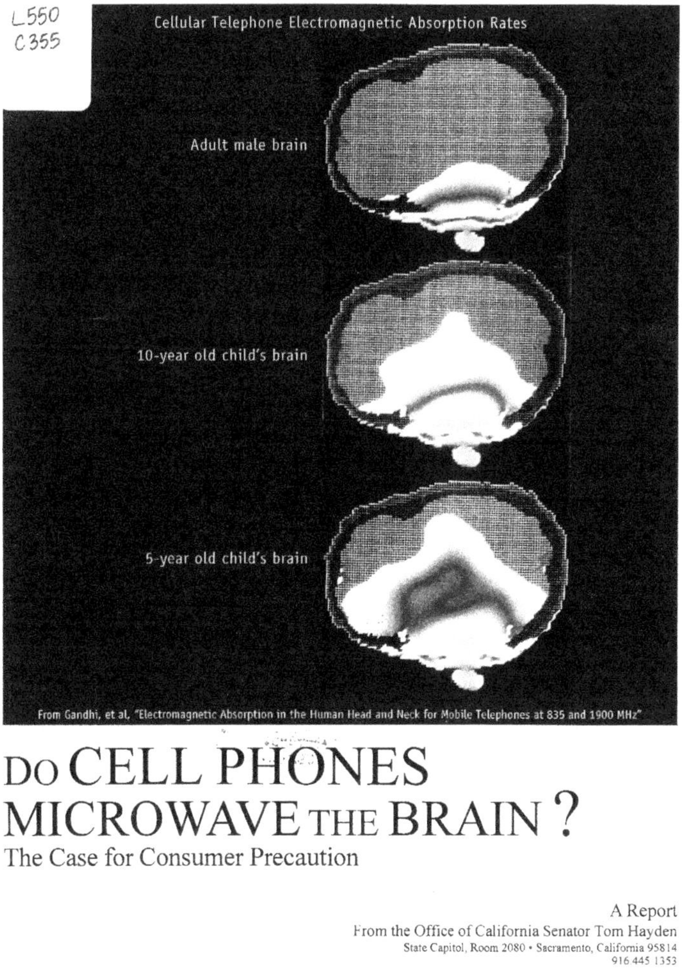

Figure 5.1. *"Do Cell Phones Microwave the Brain?"* Concerns around cell phones are one of the most familiar recent technopathologies, the subject of reports published by elected officials. Source: Book cover scan, Library of Congress. Office of California Senator Tom Hayden. Sacramento, CA: 2000.

Continuing into the early twenty-first century, consumers and activists still rallied against possible brain tumors in cell-phone users as well as various ills from the "electrosmog" of electromagnetic fields and radiation, to which cellular towers contribute.[20] This is despite research to the contrary, such as a 2009 four-country Scandinavian study and a 2010 World Health Organization 10-year study, neither of which found evidence of negative health effects from cell phone usage.[21] Yet, the technopathology persists. In 2015, the city of Berkeley, California passed a law requiring cell phone stores to inform customers of health risks.[22] The following year, *Mother Jones* reported that a "'Game-changing' study links cellphone radiation to cancer,"[23] whereas in the same year *Vox* replied, "Seriously, stop with the irresponsible reporting on cellphones and cancer."[24]

The umbrella term for these conditions related to cell phones and electromagnetism is electrosensitivity, which refers to people who suffer mental and physical ailments due to close and persistent electromagnetic radiation, even at levels that do not harm others. This hypersensitivity appears alone and in conjunction with other diseases of disputed authenticity, such as environmental illness and multiple chemical sensitivity. Its widespread, varied, and flexible nature parallels neurasthenia, and electrosensitivity similarly involves bioelectrical and central nervous system overstimulation, technological etiology, varied symptoms (e.g., headaches, muscle pains, fatigue, confusion), and frequent comorbidity with other conditions (e.g., chronic fatigue syndrome, fibromyalgia, and other environmental / autoimmune syndromes).

With electrosensitivity, we see, not only a technopathology associated with old and domesticated technologies, but the normalizing and gendering of pathologization discussed previously. And this is not directed toward the technologies; the individualization effect focuses on the user.

In the field of activism variously centered around electromagnetic radiation, electromagnetic fields, and extra-low frequencies (EMR/EMF/ELF), disease sufferers, researchers, alternative health practitioners, religious gurus, and conspiracy theorists have come together—in the type of the social movement Joseph Dumit describes as typical when organized around a liminal disease—to deal with mental and physical health problems attributed to electromagnetism.[25] They provide emotional support, give practical health advice, document suffering, organize for recognition, and protest sources of electromagnetic radiation.

I examined two types of online communication networks of electrosensitivity activism: static websites and a discussion group. The four websites were the EMR Network, the California Electric and Magnetic Fields Program, the San Francisco Neighborhood Antenna-Free Union (SNAFU), and the Council on Wireless Technology Impacts (CWTI). Every local page of these websites was examined, including PDFs and pop-up windows, for a total of 107 pages. These sites provided activist resources, legislative alerts, and education, and were easily found online. However, seeming somewhat contrary to the community-building nature of grassroots projects, they employed one-way rather than interactive communication. None offered working message boards, chat rooms, blogs, or other interactive space for activists or sufferers of EMR-related illnesses.

I also examined the archives of an open, unmoderated mailing list, unaffiliated with any organization. The list dealt with health concerns related to cell phones and towers, microwave secret weapons, power lines, and general electrosmog from electromagnetic fields and radio frequency radiation, implicating radio and TV signals, wi-fi networks, and even household appliances. The group was organized around multidirectional, asynchronous communication. Posts were unmoderated, but group policing of inappropriate posts did occur. Members often linked to external sites, but did not post static pages, files, or photographs. It had a steady posting of messages since its formation in 2000, with a total of 1874 posts by 317 members when I studied the list. I examined a systematic random sampling of 10% of all posts, excluding spam and quoted or forwarded messages.

In comparing discourses on these public websites versus the mailing list, I found that the pathologizing of the feminine persisted in the silencing of sufferers. That is, "feminine" attributes, such as sensitivity, emotion, or empathy were largely absent from the public faces of activist groups, for, as members of the mailing list would suggest, they were deemed threatening to the legitimacy of the movement. These voices were relegated to less-public spheres, demonstrating the normalizing and gendering work of pathologization. Moreover, unlike neurasthenia, this was not merely a top-down flow from institutions, elites, and authorities, but persisted even in activist groups—a legitimizing hierarchy confining certain types of knowledge, rhetoric, action, and experience to less-visible realms. Among online networks of electrosensitivity activists, the feminine face of the disease was presumed to be harmful to the broader social movement, for it raised specters of the pathological feminine: hysterical, hypochondriac, overly emotional, paranoid, and deluded. These voices and experiences were confined and relegated, much like Charlotte Perkins Gilman and her heroine(s), for exhibiting characteristics of excessive emotion and suffering, much like Louise in *Possessed*. This was explicitly directed, as in warnings from members of the discussion group, but also structural, as when their public websites' pointed lack of opportunities for individual posting or commenting preemptively prevented such individual voices of suffering from contributing at all.

Electrosensitivity, with echoes of neurasthenia, persists in popular culture as well. A 1992 Eddie Murphy film, *The Distinguished Gentlemen*, presented a cancer cluster caused by a children's playground located underneath "high-voltage power lines." As the mother of a child with cancer explained, "The wires cause magnetic fields, and the magnetic fields cause cancer, especially in children." Alan Lightman's 2000 novel, *The Diagnosis*—a National Book Award finalist—concerns a harried businessman who succumbs to overwhelming fatigue, confusion, and amnesia, wandering the city in a state, not unlike Louise in *Possessed*, and the etiology is explicitly linked to the surrounding environment of electric communication devices and technologies. Although Lightman changes the sickened protagonist to male, he maintains many attributes gendered feminine, and pathologically so. His protagonist's sickness parallels decreasing masculinity, as his wife grows shrewish, has a cyber affair, and becomes more of a drunk. He is cuckolded and further emasculated by losing his job and failing as a father. The cause of his condition is never named, but clearly mirrors the technopathology of electrosensitivity. A character on the cable TV series *Better Call Saul* lives as a recluse in his de-electrified home, only venturing outside in a foil wrap, due to his electrosensitivity.

Although electrosensitivity in its very name is more suggestive of accusing a social system—the "electro" technological infrastructure of industrialized societies—this is immediately shifted by "sensitivity." The sufferer is too sensitive—the electromagnetic radiation is not too strong or ubiquitous. In these cases, technopathologies exhibit stronger personal pathologization than many traditional diseases. When a person gets food poisoning or the flu, they get sick because of it. Although certain classes (the old, the already sick) may be more vulnerable, we generally don't have a separate class of food-poisoning-sensitives, or people who are essentially and inherently more susceptible, particularly due to inherent degeneracy or moral shortcomings. (One obvious exception being AIDS and other diseases of compromised immune systems.) The sufferers of electrosensitivity, similar to the broader pessimism encountered in my focus groups, saw systemic issues as simply too broad to face, with manufacturers and utilities impervious to change. The obvious parallel of a ubiquitous social system changing from pervasive, popular enjoyment to a widely recognized and regulated health risk—tobacco—does not seem to inspire much hope.

Common solutions are individualistic and often involving consumption (thereby clearly classed): move away to a rural cabin, measure and rehab one's home with shielding, or invest in protective cell-phone shield and EM-detecting watches. Online activists sequestered in the mailing list were those who were most willing to accuse and address systemic issues, such as overall use of technology by society, corrupt scientists bought out by business, and mind control by governments and secret elites.

While my focus group participants initially ridiculed the concept of electrosensitivity, eventually they confessed to experiencing similar technology-related health concerns. What was consistent was that the parties least responsible were systems and institutions, such as device manufacturers and government regulatory agencies. The least feasible solutions were activism, litigation, legislation, and other forms of large social change. The most common and acceptable solutions were individual changes in use—minimization and moderation of use, exercise, and other self-disciplinary practices, or purchase of protective devices or moving to safer environments. What became apparent was that electrosensitives were not sick because of their health problems, they were sick because they complained about them, and complained too loudly, and complained about systemic issues rather than personal problems of technology mis-use.

Even in movements as humanistic as alternative health and countercultural spirituality, such as kundalini yoga, the focus is often on the individual. Electrosensitivity and related syndromes, such as chronic fatigue, are blamed directly on our technological environment by leading gurus.[26] Although kundalini yoga does not pathologize the user as inherently weak or of poor moral/physical fiber, the solution is, however, wholly individualistic, consistent with the self-improvement appeal of the practice and its logic within a larger neoliberal ethos of entrepreneurial self-discipline. Environmental toxicity is more the problem than the weak individual, but the solution is still an individual's defensive strengthening and rebalancing of vital energies. Consider, by contrast, if the solution to Love Canal had been, not EPA cleanup, but somehow residents there exercising or taking supplements to have greater tolerance to industrial toxic waste, or if the anti-nuclear movement had been focused on scientific research to make people more resistant to nuclear radiation.

Pathologizing Users: The Individualization Effect

A derisive attitude toward sufferers of technopathologies often appeared in focus groups I conducted on the subject. They were described as both immature and elderly, mentally ill, irresponsible, scapegoating, "lazy," "stupid," "conspiracy theorists," "obsessive compulsive," "hypochondriacs," and needing "a rubber suit and a tin foil hat." Such attitudes suggest the individualization effect of pathologization. This is a common aspect of pathological discourses: sickness and health are viewed as problems of individuals, not societies, industries, devices, or social practices. Technopathologies are about bad users with individual problems. When talking about phenomena using patterns of pathology, the phenomenon is made legible along individualistic lines, rather than social or systemic.

Michel Foucault traces this process back to the Enlightenment. During that time, he describes a reorganization of what can be medically seen and said, of how elements of the phenomena of pathology are reconceived and redefined such that the ontology of disease moves from an outside agent to the individual itself. Foucault outlines "a welding of the disease onto the organism," arguing that, in the early 1800s, "the *being* of the disease disappears." Disease changes from a thing, an essence, to an action—that action being the patient's response to something in their environment. Carl Jung similarly defines neurosis as the response of a healthy person to their environment, even if it is an inferior form of adaptation, it can still be considered the response of a healthy organism to the demands of life.[27] It is a shift from disease as invasion or attack to disease as reaction: "Disease is now no more than a certain complex movement of tissues in reaction to an irritating cause: it is in this that the whole essence of the pathological lies, for there are no longer either essential diseases or essences of diseases."[28] This creates a "pathological responsibility of the subject for his own illness," which becomes increasingly focused on sexuality as sexual responsibility is also extended from venereal diseases to all illnesses.[29]

Such individualization is not new. Historian Andreas Killen finds a similar shift in 1926 Germany, when insurance companies stopped covering nervous disorders, making individual suffers solely to blame, and not modern society, as previously thought.[30] Link and Phelan note decades of epidemiological studies that emphasize individual risk factors (diet, exercise, lifestyle choices, etc.) and not social ones (socioeconomic status, social support), at the expense of preventative strategies that could address fundamental social causes of disease.[31] Foucauldean scholar of risk and insurance, François Ewald, writes that the increasing understanding of the source of illness as genetics could be a cultural shift toward individual responsibility, one that

> spells the end of the vision of illness inherited from Pasteur: that of an event caused by aggressive outside agents, against which the organism defends itself. In its place, we are presented with the notion that each of us is the source of our own health, that

> we are not handed the same deck of cards by nature, and that there is nothing we can do about it. Pasteur's vision was more comfortable: it meant we were not personally responsible, it considered us to be victims of illness, even though, by means of an emphasis on hygiene, it made public health a collective responsibility. The genetic vision makes us responsible for our illness: powerless, but responsible nevertheless, since we are the source of our own illness. And even more so in the sense that it makes us responsible for managing out health optimally.

Such cases suggest continuing expectations for users to take responsibility for their own pathological prevention and cures, especially as the preventative culture of risk society shifted responsibility from a moral realm to one of "just business" economics.[32] Similar perspectives, such as personally adjusting body postures to prevent repetitive stress injuries, were repeatedly asserted in my focus groups.

The individualization effect spawns two dimensions of response. One is change in individual behavior: A good user takes responsibility for changing their practices to avoid such threats. For example, online predator narratives add to discourses on appropriate user practices and ideal users. And, indeed, online sexual-predator-paranoia has been found to lead directly to individuals making changes in their technology use. A European Union study of youth and such online risks concluded that young users perceived the responsibility for safe identity management online as resting with themselves, not government, family, or industry.[33] The second dimension of response is monitoring and control of those bad users who do not amend their ways. Bad users raise questions of power and control: how will we limit them and protect others from them? Media scholar Wendy Chun argues that, drawing on Foucault's analysis of sexuality as locus of subjectivity and power relations, as communication technologies threaten established orders, "the relationship between control and freedom … is often experienced as sexuality or is mapped in terms of sexuality-paranoia."[34] Indeed, the rapacious cultural appetite for narratives of online predators caught and punished demonstrates this.

Predators

Today, perhaps one of the most familiar technological connection to dangers would be online sexual predators. These are technopathologies in the sense of the technology being a carrier that connects a user to a threat to their health and well-being. The technology does not cause the harm, but is a necessary condition for it, seen in films such as *Trust*, showcased exhaustively on the entrapment television series, *To Catch a Predator*, and when *Vanity Fair*, in 2009, claimed that the Internet "made possible" and "enabled" the crimes of the "Craigslist Murderer,"[35] later dramatized in the film *The Craigslist Killer*. New media scholar Alice Marwick has argued that the inflated sense of threat, disproportionate to the actual statistics for online predation by strangers (particularly when compared to the statistics for offline predation by family and acquaintances) constitutes a moral panic. Marwick refines this as a *technopanic*, which she describes as a "moral panic as a response to fear of modernity as represented by new technologies."[36] Similarly, Sarah Banet-Weiser has analyzed fears around the Internet and related technologies as displaced anxieties around new degrees of lost parental control and family structural changes.[37]

However, such scholarship, while insightful and in many ways accurate, reaffirms the newness hypothesis and potentially contributes to a misplaced focus on technologies over users, which this book seeks to redress. Even though the technologies here are not to blame, they still remain the center focus of discourses—what Slack and Wise call the "star-struck actor center stage" of technological determinism, the blank slate, the innocent victim upon which social anxieties are misplaced.[38] Marwick's term is "*techno*panic," after all. Moreover, while elevating the focus on technology, real people are pathologized as the anxious or otherwise irrational users unable to recognize the true source of their fears. It is for this reason that I purposefully avoid reducing technopathologies to solely metaphors or displacements, or engaging in the question of which technopathologies are "real." For, even in the case of technopanics, real users are really panicked.

But predator technopathologies are not about new media. The first commercial microcomputer with mouse and keyboard dates back to 1959, with Digital Equipment's Programmed Data Processor 1. Popular media coverage and adoption of personal computers occurred in the 1980s, and the widespread adoption of the Internet took place in the 1990s. Early popular social networking sites included Geocities in 1994, Six Degrees in 1997, and Friendster in 2002.

Predator technopathologies are about users. In cinematic thrillers about techno-snuff, such as *.Com for Murder*, *8mm*, *Carver*, *Dead and Buried*, *Death Tube*, *Death Tube 2*, *Dot.Kill*, *Feardotcom*, *Feed*, *Killer Movie*, *Reality Kills*, *Reincarnation*, *Rest Stop*, *Sinister*, *Tesis*, *Untraceable*, *Vacancy*, *Vacancy 2*, *Vlog*, and *WatchUsDie.com*, communication technologies help locate, stalk, and/or record or broadcast the murders of their victims. Slater and Kerekes argue that, in such films, the communication technologies *make things worse*: they amplify the murder's pathology, similar to how a crime, injury, or sickness becomes more newsworthy when involving communication technologies.[39] In other words, although the technology is central to these narratives, it is not the technology itself that it pathologized: it is the sick(ening) user to whom technologies expose us.

Moreover, these predator panics are not so new. In the mid-1800s, a New England professor weighed concerns about the risks of future telegraph cables linking Americans to possibly diseased and dangerous Europeans.[40] In the 1880s, as Charleston, South Carolina, was preparing for installation of electrical wires and support poles, some worried that the poles would provide "excellent places for persons wishing to assault one another to stand behind."[41] Elsewhere, a social reformer worried about the telegraph, "'It is a well-known fact that no other section of the population avail themselves more readily and speedily of the latest triumphs of science than the criminal class'."[42]

Likewise, an ongoing concern has been the ability of telephones to act as a carrier that exposes users to sickening, pathological users who, in turn, generate fear, anxiety and other forms of mental distress. In the 1902 Grand Guignol play by André de Lorde, *At the Telephone*, a psychopath forces a man to listen to the murder of his wife and child over the phone.[43] Decades later, a tabloid news article from 1969 began, "Sick minds are using the nation's telephone network to torment innocent people." It continued to relate that a psychiatrist described perpetrators as "telephone tormentors … exhibitionists … some are schizophrenics or split personalities. … Phone calls from strange men can drive a woman to hysteria. Model (above) shows how many helpless victims react to obscene calls."[44] Note here again the relationship between pathologization and gendering: female recipients are apparently "helpless" to hang up or change their phone numbers. On one hand, it is a positive sign that a situational context is acknowledged for upsetting women (rather than merely it being their own fault for being oversensitive neurotics). However, this is contained by the sickness to which they are driven: hysteria, a diagnosis that has historically performed cultural work of blaming the feminine victim. The user could simply hang up, but bad users persist in answering the phone, engaging in discussions with the harasser. They are naïve, foolish, unsavvy users. And why do men so rarely receive obscene phone calls? Furthermore, although the male perpetrators are at times indicted ("torment(ors) … exhibitionists … abusive … strange"), there are also medical suggestions of mental illness that could possibly excuse them ("schizophrenics or split personalities").[45]

Hysteria and other forms of distress caused by telephonic exposure to psychopaths have demonstrated such persistence as to become a cinematic cliché, one that continues well into decades of mobile telephony. The trope appears frequently in horror films and thrillers.[46] As one film scholar has argued, the telephone acts not merely as an innocent carrier that connects users to pathological persons, it is almost an autoimmune disorder in that it lowers users' defenses and makes them vulnerable: "Even when the telephone is still operative one remains vulnerable; in fact it is the telephone that makes us vulnerable. … You have the killer using the telephone as a means to violate the victim."[47]

Predator technopathologies continued into early forms of computer networking. In the late 1960s, the *National Enquirer* warned that "Rapists and sexual perverts are using computerized dating services to line up unsuspecting girls as potential victims" in the lede to a story headlined, "Male Misfits Use Computer Dating Services to Meet Innocent Girls."[48] In 1997, the Internet was blamed in "Kids Who Kill," a 4-page spread in the *National Enquirer* that warned of murderous teens online, such as a sadistic "computer nerd," "unsavory losers who warp young minds," and other "violence-prone kids" that are "hooked on the Internet," which is "an electronic sewer, home to slime who prey on the young and vulnerable." The article promises, in an upcoming issue, a special investigation into "Sick Monsters Stalking Your Child on Internet."[49]

In my focus groups, one participant warned of a very different way technology could connect naïve users to dangerous degenerates: a car's GPS navigation systems had been used to connect car thieves to a driver's home, which they could then rob. (Although, another participant pointed out that the driver's address is also typically on their registration in the glove compartment).

Online predators are clearly pathological, but the bad user who promiscuously connects to them is far from innocent. Young women revealing too much of themselves online—parallel to revealing too much of their bodies when dressed "provocatively"—can attract a killer's attention, in films such as *The Den* and *Vlog*. Like a rape victim blamed for walking alone at night or wearing tight clothing, threats of online predators easily shift to discourses of proper technological behavior for those who do not wish to be their victims. For example, fatally hooking up with degenerates online is the central conceit of Rick Reed's novel *IM*. Taglined "Instant Message or Instant Murder," this gay erotic mystery concerns the use of adult social networking sites for casual sex, and it is technology that exposes users to a homicidal sociopath. In *IM*, the pathologized usership of online cruising is reinforced by the pathologizing of the killer—but also his victims—as bad gay role models. The killer is feminine and girlish, and the victims exhibit drug and alcohol abuse, unsafe sex, superficiality, vanity, and promiscuous hooking up online—all in contrast to the masculine, monogamous, and sober protagonist, who doesn't even use online adult services. When I interviewed Reed, who frequently has technology themes in his work, he denied consciously making any overt morality tale about what happens to bad users online. However, he did describe the moral of the book as one of responsible technological usership:

> The Internet can often be a lot of smoke and mirrors and even if you think you know with whom you're hooking up, use caution. Meet first in a public place. Tell someone you trust where you're going if you're meeting up with someone. There are no guarantees for either bad or good resulting from Internet interaction, but there are precautions that might help tip the scales in your favor.

The cyberbully figure represents a most recent incarnation of the degenerate online predator threat, the topic of numerous news specials and fictional accounts, ranging from the PBS Frontline documentaries "Growing Up Online" and "Generation Like" to the movies *#Horror* and *Chatroom*.

Technopathology and Common Users

As I have suggested, pathologizing individual users is not limited to technologies that are new, perceived as new, or even particularly spectacular. As with electrification in Anderson, these processes can work at quite a quotidian level, with the individualization effect pathologizing the users of technologies as mundane and commonplace as office printers.

"Laser Printers Can Kill You with Evil Dust" blared a sarcastic headline in *Gadget Lab*, part of the blog network of *WIRED* magazine. The post described how Australian researchers had found that some laser printers emitted small particles that could lodge in pulmonary crevices, resulting in cardiovascular difficulties, respiratory irritation, and cancer. As the hyperbolic headline suggested, the post expressed an irreverent attitude toward the study's findings. The *WIRED* blogger described the study as proclaiming laser printers "could be as dangerous as cigarettes," a health concern it dismissed as "another scare story along the lines of WiFi frying children's brains."[50]

Moreover, the target of the post's derision moved from the research findings to persons who might take seriously the idea of people sickened after using laser printers. Its conclusion indicted the lead researcher: "The professor in charge of the study is already calling for nannyfication of the problem: 'Governments regulate emission levels from outdoor devices such as vehicles, power stations and factories,' said Professor Lidia Morawska, 'so why not for printers?' Indeed. And ban the machines that make scalding hot coffee while you're at it."

A discussion of potential health risks had quickly progressed into personal attack. The post alludes to an infamous case in which a McDonald's customer—an elderly woman—sued the fast-food chain for burns suffered when she spilled their hot coffee on herself. Colloquially, the case has become a paradigmatic anecdote to describe elements of American society as over-litigious, relying on the money from lawsuits for income rather than meritorious hard work, rhetorically parallel to, for example, the epithet "welfare queens." In so doing, the blogger took a story of a potentially dangerous *technology* and reframed it as one of bad technology *users*: those who irrationally fear technology (wi-fi paranoia), and, in cahoots with alarmists and condescending over-regulators, go against the libertarian strain of techno-utopianism that *WIRED* exemplifies. The technology misuse is also, by implication, a direct threat to "normal" technology users. A case of bad technology use is linked to bad consumerism: The logic—a familiar one in U.S. political discourse for decades—is that belief in laser printer illnesses could lead to lawsuits and, therefore, passed-on costs to all consumers as well as increased legislation. This, in turn, stifles innovation, depriving others of new technological innovations, hurts the national economy, costs jobs, and cripples U.S. global competitiveness. In other words, bad user-consumers are not merely worthy of derision, they are a direct material threat to the livelihoods of individuals and the health of the state.

The McDonald's coffee lawsuit is often used as evidence of consumer stupidity abetted by legal greed and corruption in support of tort reform legislation. The blogger makes what could be a problem of motor skills, balance, beverage temperature, container design, customer service, or menu planning and implementation into one about technology: "ban the *machines* that make scalding hot coffee." The post also rhetorically reinforces the stereotypical association of technological ineptitude with femininity and advanced age. It makes a series of leaps from the laser printer study's sober call for more investigation into a heretofore-unstudied phenomena (particle emissions), to a decades-long struggle of research, consumer education, litigation, and corporate malfeasance involving a leading cause of death (smoking), to what is generally thought to be a frivolous lawsuit by an idiot.

The blogger perhaps can be excused, for he relied on, not the actual unpublished study, but news reports of it.[51] However, this does not lessen the cultural work such articles can perform, informing conceptions, ideals, and expectations of technology use and users. The blog post not only ridiculed the study, but also reinforced through repetition the conventional wisdom regarding the coffee lawsuit, its status as common knowledge, or ideology: something everyone knows.

Except, typically, they don't. Such anecdotal mentions of the coffee lawsuit typically fail to acknowledge that the woman suffered third-degree burns over 6% of her skin and lesser burns over 16%, affecting her groin, thighs, and buttocks, and requiring eight days of hospitalization for skin grafts and two years of follow-up treatments. She initially sought $20,000 to cover medical costs of $11,000, to which McDonald's offered $800, refusing to raise its offer or settle. This led to the infamous lawsuit and a $2.86 million jury award (a figure based on a percentage of McDonald's profits from coffee sales), reduced by the judge to $640,000, then settled out of court during appeals for less than $600,000. At the time, McDonald's required drive-through coffee to be served at 180–190 degrees, had already received over 700 complaints regarding burns and scalding, and had already awarded settlements of up to $500,000.[52]

Within the context of *WIRED*, the implication seems that technological health concerns are at odds with progress. The digerati don't get sick. As suggested by the laser printer story, concerns about technology and illness tend to not demonize sickening technologies, but pathologize sick users and, by extension, those who take seriously sick users.

Notes

1. Acland, "Introduction"; Chun, "Introduction"; Gitelman and Pingree, *New Media*; Marvin, *Old Technologies*; Gitelman, *Always Already New*.
2. Stern, "Addiction."
3. Stern and Handel, "Sexuality."
4. Simon, *Dark Light*, 145, 10.
5. Shotton, *Computer Addiction*, 235.
6. Ghorayshi, "Google Glass."
7. Konnikova, "Is Internet Addiction a Real Thing?".
8. Marvin, *Old Technologies*, 117, 132–133, 151.
9. Spigel, *Make Room*, 4.
10. O'Neill, "Lose Weight?"
11. Brox, *Brilliant*.

12. Bach, "Glass Arm."
13. Lanza in *Elevator Music* suggests this trend's origins in the Industrial Revolution, with conditions such as "Boilermaker's Disease" brought on by factory noise, such as scraping metal.
14. "Notes and News."
15. Coz, "Headaches."
16. Stern and Gwathmey, *Telephone*, 70.
17. Becker and Selden, *Body Electric*.
18. Hayden, *Cell Phones*.
19. Burgess, *Cellular Phones*, 25.
20. Burgess, *Cellular Phones*; Carlo and Schram, *Cell Phones*; Hayden, *Cell Phones*; Kane, *Cellular Telephone*; Leonardi, "Problematizing."
21. Szabo, "Four-Country Study"; Szabo and Marcus, "WHO". Yet the WHO study still deemed the phenomena "inconclusive" because it could not entirely rule out the possibility of potential health hazards from cell phones. This evidences what Burgess calls a "culture of precaution," an insistence that health fears persist until something can be proved 100% safe; a standard to which we do not submit other risks, such as those we know to be potentially unsafe but accept the risk (e.g., driving in automobiles) or the fact that you cannot disprove a potential. Less than two years prior one co-author had written a story about a four-country study finding no cancer link, but didn't mention this in the 2011 article.
22. "Cell Phones."
23. Harkinson, "Study."
24. Plumer, "Stop."
25. Dumit "'Come on'", "Illnesses."
26. Shannahoff-Khalsa, *Kundalini*, 183–190.
27. Jung, *Collected Works*.
28. Both Foucault, *Birth of the Clinic*, xix.
29. Foucault, "5 March 1975," 241–242.
30. Killen, *Berlin*.
31. Link and Phelan, "Social Conditions."
32. Ewald, "Genetics," 18, "Return."
33. Lusoli and Miltgen, *Young People*.
34. Chun, *Control*, 11.
35. Orth, "Killer@Craigslist," 156.
36. Marwick, "Predator," "Technopanics" section, ¶1.
37. Banet-Weiser, "Surfin' the Net."
38. Slack and Wise, *Technology*.
39. Slater and Kerekes, *Killing for Culture*.
40. Marvin, *Old Technologies*, 132.
41. Pogue, *South Carolina*, 11.
42. Quoted in Standage, *Victorian Internet*, 105.
43. de Lorde, *Telephone*.
44. Blaine, "Obscene and Threatening," 25.

45. Furthermore, this is an excellent example of the very semantic confusion around the most common term for illness: the first word of the article, "sick." Although medically and culturally much work has been done to differentiate between cruel criminality and innocent mental illness, both meanings persist in the single word: *That was a sick thing that he did, but he can't help it because he's sick.* The slang adoption of the "sick" by youth and "sickening" by drag queens, both to mean intensely impressive, adds another layer of complexity. Such semantic confusion suggests the conceptual confusion that thousands of courtroom dramas have yet to clarify.
46. *Are You in the House Alone?*, *Black Christmas* (1975, 2006), *The Caller*, *Don't Hang Up*, *Murder-Rock Dancing Death*, *Message in a Cell Phone*, *Scream*, *Scream 2*, *Scream 3*, *Scream 4*; *Silent Night, Bloody Night*; *I Still Know What You Did Last Summer*, *Long Distance*, *When a Stranger Calls* (1979, 2006), and *Zodiac*. In *Vlog*, voicemails direct a video blogger to see video clip postings of subjects from her blog being murdered. In *Joy Ride* and *Joy Ride 2: Dead Ahead*, the killer harasses by CB radio—hardly a new medium in the twenty-first century.
47. Spiteri, "Isolation" section, ¶ 3.
48. Mahoney, "Male Misfits."
49. Blosser and Nelson, "Kids Who Kill," 24–27.
50. Sorrel, "Laser Printers," 1.
51. "Laser Printers;" Judge, "Office Printer."
52. "FAQ About."

References

Acland, Charles R. "Introduction: Residual Media." In *Residual Media*, edited by Charles R. Acland, University of Minnesota Press, 2007, pp. xiii–xvii.

Are You in the House Alone? Dir. Walter Grauman. 1978, USA.

Bach, J. K. "The Glass Arm … the Dreaded CW Man's Disease." *The Master Wood Butcher's Glass Arm Page*, 2012. http://www.woodbutcher.net/glass-arm.shtml. Accessed May 2, 2013.

Banet-Weiser, Sarah. "Surfin' the Net: Children, Parental Obsolescence, and Citizenship." In *Technological Visions: The Hopes and Fears that Shape New Technologies*, edited by Doug Thomas, Marita Sturken, and Sandra J. Ball-Rokeach, Temple University Press, 2004.

Becker, Robert, and Gary Selden. *The Body Electric: Electromagnetism and the Foundation of Life*, Morrow, 1985.

Black Christmas. Dir. R. Clark. 1975, USA.

Black Christmas. Dir. G. Morgan. 2006, USA.

Blaine, W. "Obscene and Threatening Phone Calls Are Increasing at an Alarming Rate." *National Enquirer*, 1969, February 16, p. 25.

Blosser, J., and J. Nelson. "Kids Who Kill." *National Enquirer*, 1997, October 21, pp. 24–27.

Brox, Jane. *Brilliant: The Evolution of Artificial Light*, Houghton Mifflin, 2010.

Burgess, Anthony. *Cellular Phones, Public Fears and a Culture of Precaution*, Cambridge University Press, 2004.

The Caller. Dir. Matthew Parkhill. 2011, USA.

Carlo, G., and M. Schram. *Cell Phones: Invisible Hazards in the Wireless Age*, Carroll and Graf, 2001.

"Cell Phones and Risk of Brain Tumors—What's the Real Science?" *CNN Wire*, 2015, July 28. http://fox61.com/2015/07/28/cell-phones-and-risk-of-brain-tumors-whats-the-real-science/. Accessed July 14, 2017.

Chun, Wendy Hui Kyong. *Control and Freedom: Power and Paranoia in the Age of Fiber Optics*, The MIT Press, 2006.

Chun, Wendy Hui Kyong. "Introduction: Did Somebody Say New Media?." In *New Media, Old Media: A History and Theory Reader*, edited by Wendy Hui Kyong Chun and Thomas Keenan, Routledge, 2006, pp. 1–10.

Coz, Steve. "Easy Way to Avoid Headaches—By Cutting Out These Bad Habits." *National Enquirer*, 1983, August 2, p. 51.

de Lorde, Audre. "At the Telephone." In *One-act Plays for Stage and Study, second series [Electronic version]*, edited by W. P. Eaton, Gaslight/Samuel French, 1925.

Don't Hang Up. Dir. S. F. Brownrigg. 1974, USA.

Dumit, Joseph. "'Come On, People … We *Are* the Aliens. We Seem to be Suffering from Host-Planet Rejection Syndrome': Liminal Illness, Structural Damnation, and Social Creativity." In *E. T. Culture: Anthropology in Outerspaces*, edited by D. Battaglia, Duke University Press, 2005.

Dumit, Joseph. "Illnesses You have to Fight to Get: Facts as Forces in Uncertain, Emergent Illnesses." *Social Science and Medicine*, no. 62, 2006, pp. 577–590.

Ewald, Francois. "Genetics, Insurance, and Risk." In *Genetics and Insurance*, edited by U. Wiesling, T. McGleenan, and Francois Ewald, Springer-Verlag, 1999.

Ewald, Francois. "The Return of Descartes's Malicious Demon: An Outline of a Philosophy of Precaution." In *Embracing Risk: The Changing Culture of Insurance and Responsibility*, edited by Tom Baker and Jonathan Simon, University of Chicago Press, 2002, pp. 273–301.

"FAQ About the McDonald's Coffee Case and the Use of Fabricated Anecdotes." *Center for Justice and Democracy at New York Law School*. http://centerjd.org/content/faq-about-mcdonald's-coffee-case-and-use-fabricated-anecdotes. Accessed June 9, 2013.

Foucault, Michel. *The Birth of the Clinic: An Archaeology of Medical Perception*, Vintage, 1994.

Foucault, Michel. "5 March 1975." In *"Abnormal": Lectures at the Collége de France, 1974–1975*, edited by François Ewald, M. Bertani, and A. Fontana, New York: Picador, 1999.

Ghorayshi, Azeen. "Google Glass User Treated for Internet Addiction Caused by the Device." *The Guardian*, 2014, October 14. https://www.theguardian.com/science/2014/oct/14/google-glass-user-treated-addiction-withdrawal-symptoms. Accessed July 16, 2015.

Gitelman, Lisa. *Always Already New: Media, History, and the Data of Culture*, The MIT Press, 2006.

Gitelman, Lisa, and Geoffrey B. Pingree, eds. *New Media, 1740–1915*, The MIT Press, 2003.

Harkinson, Josh. "'Game-changing' Study Links Cellphone Radiation to Cancer." *Mother Jones* 2017, June 12. http://www.motherjones.com/environment/2016/05/federal-study-links-cell-phone-radiation-cancer/?google_editors_picks=true. Accessed June 12, 2017.

Hayden, Tom. *Do Cell Phones Microwave the Brain? The Case for Consumer Precaution*. Office of California Senator Tom Hayden, 2000.

I Still Know What You Did Last Summer. Dir. D. Cannon. 1998, USA.

Joy Ride. Dir. John Dahl. 2001, USA.

Joy Ride Ahead: Dead Ahead. Dir. Louis Morneau. 2008, USA.

Judge, P. "Is Your Office Printer Giving You the Sniffles?" *Silicon.com*. 2007. http://management.silicon.com/itpro/0,39024675,39168025,00.htm. Accessed on May 19, 2010.

Jung, Carl. *The Collected Works of C. G. Jung Vol. XIII: Alchemical Studies*, Princeton University Press, 1967.

Kane, R. C. *Cellular Telephone Russian Roulette: A Historical and Scientific Perspective*, Vantage Press, 2001.

Killen, Andreas. *Berlin Electropolis: Shock, Nerves and German Modernity*, University of California Press, 2006.

Konnikova, Maria. "Is Internet Addiction a Real Thing?" *The New Yorker*, 2014, November 26. http://www.newyorker.com/science/maria-konnikova/internet-addiction-real-thing. Accessed November 27, 2014.

Lanza, Joseph. *Elevator Music: A Surreal History of Muzak®, Easy-listening and Other Moodsong®*, revised and expanded edition, University of Michigan Press, 2004.

"Laser Printers May Pose Health Risks." *ZDNet*. 2007. http://news.zdnet.co.uk/hardware/0,1000000091,39288337,00.htm. Accessed August 1, 2007.

Leonardi, M. "Problematizing 'New Media': Culturally Based Perceptions of Cell Phones, Computers, and the Internet among United States Latinos." *Critical Studies in Mass Communication*, vol. 20, no. 2, 2003, pp. 160–179.

Link, Bruce G., and Jo Phelan. "Social Conditions as Fundamental Causes of Disease." *Journal of Health and Social Behavior*, no. 35 (extra issue), 1995, pp. 80–94.

Long Distance. Dir. M. Stern. 2005, USA.

Lusoli, C., and W. Miltgen. *Young People and Emerging Digital Services: An Exploratory Survey on Motivations, Perceptions and Acceptance of Risks*, European Union European Commission Joint Research Center; Institute for Prospective Technological Studies, 2009.

Mahoney, R. "Male Misfits Use Computer Dating Services to Meet Innocent Girls." *National Enquirer*, 1969, February 16, p. 21.

Marvin, Carolyn. *When Old Technologies were New: Thinking about Electric Communication in the Late Nineteenth Century*, Oxford University Press, 1988.

Marwick, Alice. To Catch a Predator? The MySpace Moral Panic. *First Monday 13*, No. 6, 2008. http://firstmonday.org/ojs/index.php/fm/article/view/2152/1966. Accessed June 6, 2013.

Message in a Cell Phone. Dir. Eric Hendershot. 2000, USA.

Murder-Rock Dancing Death. Dir. L. Fulci. 1984, Italy.

"Notes and News." *Pastoral Psychology*, vol. 16, no. 8, 1965, pp. 53–55.

O'Neill, William. "Want to Lose Weight? Turn Down the Lights." *National Enquirer*, 1983, July 19.

Orth, Maureen. "Killer@Craigslist." *Vanity Fair*, 2009, October, pp. 156–163.

Plumer, Brad. "Seriously, Stop with the Irresponsible Reporting on Cellphones and Cancer." *Vox*. 2016, May 27. https://www.vox.com/2016/5/27/11797924/cellphones-cancer-bad-reporting. Accessed July 23, 2017.
Pogue, Nell Coleman. *South Carolina Electric & Gas Company, 1846–1964*, South Carolina Electric and Gas Company, 1964.
Scream. Dir. W. Craven. 1996, USA.
Scream 2. Dir. Wes Craven. 1997, USA.
Scream 3. Dir. Wes Craven. 2000, USA.
Scream 4. Dir. Wes Craven. 2012, USA.
Shannahoff-Khalsa, D. S. *Kundalini Yoga Meditation: Techniques Specific for Psychiatric Disorders, Couples Therapy, and Personal Growth*, W. W. Norton, 2006.
Shotton, M. *Computer Addiction? A Study of Computer Dependency*, Taylor & Francis, 1989.
Silent Night, Bloody Night. Dir. Theodore Gershuny. 1972, USA.
Simon, Linda. *Dark Light: Electricity and Anxiety from the Telegraph to the X-ray*, Harcourt, 2004.
Slack, Jennifer D., and J. MacGregor Wise. *Culture + Technology: A Primer*, Peter Lang, 2005.
Slater, D., and D. Kerekes. *Killing for Culture: An Illustrated History of Death Film from Mondo to Snuff*, Creation Books, 1995.
Sorrel, C. "Laser Printers Can Kill You with Evil Dust." *WIRED*, 2007, August. http://blog.wired.com/gadgets/2007/08/laser-printers-.html. Accessed August 1, 2007.
Spigel, Lynn. *Welcome to the Dreamhouse: Popular Media and Postwar Suburbs*, Duke University Press, 2001.
Spiteri, C. "Isolation and Subjugation: The Telephone in the Slasher Film." *Senses of Cinema*, no. 32, 2004.
Standage, Tom. *The Victorian Internet: The Remarkable Story of the Telegraph and the Nineteenth-Century's On-line Pioneers*, Walker Publishing, 2007.
Stern, Steven E. "Addiction to Technologies: A Social Psychological Perspective of Internet Addiction." *CyberPsychology & Behavior*, vol. 2, no. 5, 1999, pp. 419–424.
Stern, Steven E., and Alysia D Handel. "Sexuality and Mass Media: The Historical Context of Psychology's Reaction to Sexuality on the Internet." *The Journal of Sex Research*, vol. 38, no. 4, 2001, pp. 283–291.
Stern, Ellen, and Emily Gwathmey. *Once Upon a Telephone: An Illustrated Social History*, Harcourt Brace and Company, 1994.
Szabo, Liz. "Four-Country Study Finds No Cancer Link to Cellphone Usage." *USATODAY.com*, 2009, December 3.
Szabo, Liz, and Mary Marcus. "WHO: Cellphones Possibly Carcinogenic," *USA Today*, 2011, June, 1A.
Vlog. Dir. Joshua Butler. 2008, USA.
When a Stranger Calls. Dir. F. Walton. 1979, USA.
When a Stranger Calls. Dir. S. D. West. 2006, USA.
Zodiac. Dir. David Fincher. USA, 2007.

· 6 ·

TECHNOPATHOLOGIES AS SOCIAL DISEASE

Reproducing Good and Bad Users

"Stay here and keep watching TV until you die"

—*Dead Waves*

I have shown how pathologization is a process of constituting subjectivities, and I argued for conceiving of users as not a replacement for, but neoliberal acceleration of media audiences. I have traced the rational, deep-rooted cultural logics of pathological technoculture in the intertwined histories of science, spirituality, the paranormal, and medicine. The separation of these distinct realms of phenomena allowed for causal relationships between them, but I argue their shared histories and interconnections simultaneously provide a cultural logic for technopathologies. I have demonstrated how technopathologies are not exclusively a dystopian reaction to new technologies. Instead, I have argued, they are constitutive discourses of users and ideal usage, of integrating technology use into individual and social lives and phenomena, often connecting to existing social categories and power hierarchies. Through the individualization effect, technopathologies obscure social and systemic factors and responsibilities.

In this chapter, my argument progresses to show how, although they may individualize, technopathologies are *social* diseases. Drawing on the medical history of degeneracy, I will argue that technopathologies exhibit a eugenic logic. The first response from this logic is to stop sick users from reproducing. Evidencing an explicitly sexual concern over reproduction of bad users, the individual user becomes a reproductive threat to the larger collection of audiences or networks. Sick users are not victims but threats who can infect good, healthy users, and also reproduce more bad users. The second response from a eugenic logic is encouraging the production of good users: discourses—technopathological and otherwise—that shape ideals and expectations of usership.

Foucault notes, "The political significance of the problem of sex is due to the fact that sex is located at the point of intersection of the discipline of the body and the control of the population."[1] With users, however, we have asexual reproduction, yet the contours of the discourses are remarkably similar. Furthermore, as has been suggested previously, the nature of pathology also serves to obscure social causes and contexts, individualizing and personalizing social threats. Moreover, the social implications serve to align with and reinforce other vectors of social hierarchies.

First, I will utilize a new analytic by applying pathological analysis to a healthy, good user: midcentury actor Robert Montgomery, a television consultant and reform advocate. Here, I argue that a good, healthy audience member is conceived of as a user, even at the height of the mass-broadcasting era. In the second half of this chapter, I examine a cluster of media texts derived from Koji Suzuki's techno-horror novel *Ringu*, in which supernatural plagues travel through videotapes, telephones, and virtual worlds. Drawing on sound studies and feminist scholarship, I will reflect on technopathologies' concern with reproduction of technological subjects. The additional threat that develops with technopathologies in mass communication technologies is not only bad users who must be prevented from hurting themselves and other users, but also from making new, bad users.

Degeneracy: Social Diseases

The archaic medical condition of degeneracy involves lesser-developed and/or evolutionarily backsliding lifeforms, or the ideas or habits of such lifeforms, who infect, tempt, pervert, or otherwise drag down the normal population.[2] An example of this would be the historic stereotypes of female prostitution. According to this, women, especially lower-class and women of color, succumb to prostitution because they have less moral fortitude to resist temptation, persuasion, and earthly pleasures—including consumer enticements. Shopping and leisure entertainments, such as attending the movies, were feared as presenting women with too many options upon which to spend money, and prostitution presented a ready method for refilling pocketbooks. Then, they solicit men, corrupting their characters and spreading venereal diseases to them, and their wives and children.

Established in 1857, Foucault considers degeneracy the conceptual lynchpin in medical science's annexation of the abnormal, through linkage of insanity with the concept of the "dangerous individual." Psychiatry connected any abnormality with threatening degeneracy, by which the medical field "gained a possibility of indefinite intervention in human behavior."[3]

Degenerates were more than mere "bad apples." They were not simply sick persons who existed "out there" with the potential to infect others. Infecting others was, if not their prime directive, a direct result of many of their main characteristics. In the historic case of mulattoes, degenerate persons of mixed racial background in the nineteenth-century U.S. were considered sterile, doomed to die out as a social group or identity unless they found other ways to reproduce, suggesting a motivation for encouraging others to participate in the sorts of degenerate activities that produced them—miscegenation. Homosexuals were thought to notoriously recruit and convert. Echoing neurasthenia, degenerates were thought to be cases of nervous exhaustion, neurological overstimulation, and hypersensitivity. Their resulting disinhibition was a key diagnostic criterion: an inability to control and regulate oneself in order to appropriately integrate into society. Degenerates, then, were not accidentally antisocial, and they were inherently, actively so. Their "search for strong stimulation occasioned by abnormal nerve transmission [led] to love of the strange, bizarre, evil, loathsome, and ugly, and to sexual perversions."[4] Degenerates' egomania heightened this threatening picture: inherently sick people who craved sick stimulation and interaction didn't care about anyone who got hurt in their quest. They were, therefore, not merely potentially corrupting, but actively so. For a technological example, itinerant telegraph operators, or "boomers," were associated with many degenerate characteristics: They were thought to be alcoholic and mentally ill, antisocially withdrawing from direct human contact in preference for machines, and promoting unrestrained sexuality and dangerous flirtations through the wires.[5]

Eugenic Logic

Degeneracy's social threat leads to a logic of eugenics. As a social philosophy, eugenics was developed by Francis Dalton, cousin of Charles Darwin. From the 1900s to early 1930s, its notion of self-directed evolution became an academic discipline with university departments, international conferences, and popular societies around the world. Various eugenic policies, such as sterilization of the mentally ill or disabled, were implemented in the United States and throughout other Western nations. The logical apex of its ideology, Nazi Germany, brought about its rapid decline in popularity.

The degenerate is not merely a danger; the degenerate is an incurable danger. Foucault describes a "racism against the abnormal,"[6] in which psychiatry abandons trying to cure and, instead, focuses on protecting society, asserting a new function of "the detection of all those within a group who may be the carriers of a danger to it."[7] In the twentieth century, Foucault asserts that Nazism bonds this "racism against the abnormal" with traditional ethnic racism in their genocidal policies, and parallels can be seen in the American eugenics moment. "The degenerate is someone who is a danger. … The degenerate is someone who, at all events, cannot be cured."[8]

The logic of eugenics suggests two responses. One is what eugenics is most famous for: Stop the sick from reproducing. This is a popular strategy we see in efforts to sterilize users by disconnecting them from the network. In times of perceived threat or instability, local and national governments have increasingly dampened cellular networks, blocked specific applications, such as Twitter, or disconnected Internet access. Users can be banned from Facebook or Google for obscene or hateful postings but also have been banned for not using their real names or identities—all in striking contrast to the early days of the Internet. Businesses and public spaces also force users to disconnect, banning cell phone use or head-mounted Internet cameras due to their potential inappropriate uses. Mobile applications such as the now-defunct YikYak had to incorporate "geofencing" to block their use around schools, where hateful and bullying anonymous posts were causing disruptions.

However, this was a repressive approach, one Foucault would perhaps describe as sovereign power: those in power prevent you from doing something by force. Although certainly real, he urges us to be cognizant of what he calls modern power, ways in which we willingly participate in power relationships through a variety of daily practices, many often quite pleasurable. Technopathological discourses of being a good user, and reproducing other good users, are part of this process. Bad usage practices are not simply outlawed, new categories of healthy and sick, good and bad, users are produced. We willingly align and strive to be good users. This is the second eugenic response: discourses—technopathological and otherwise—that shape ideals and expectations of usership.

Mass-Media Technopathologies

Mass-media technopathologies elevate concerns to the social level, while continuing to nevertheless blame the individual. However, it is not only the individual sick user that is the problem, it is their reproduction of new sicknesses and new sick users. Whether the threat is television brainwashing or supernatural plagues, with the advent of mass-media broadcasting, technopathologies become elevated to social concerns. Rather than mere stigmatization or stereotyping, this raises the stakes of categorizing good and bad users. If bad users can infect others in networks or reproduce themselves across them, then they become a public health issue warranting greater monitoring, control, and regulation.

There are numerous examples of technopathologies related to broadcast mass media. As scholars such as Michelle Hilmes have shown, the connectivity of radio was threatening when it brought users—particularly those in the white middle class—in contact with the music and culture of ethnic and often African-American cultures.[9] Metaphors of infection and contagion have a long history of use in describing the spread of African-based sounds and musics.[10] Another radio scholar describes the recorded voice on radio as inherently hallucinatory and schizophrenic.[11] Headaches were attributed to 3D movies, not when they first appeared, but during their second major revival decades later, further disproving the newness hypothesis.[12] Similarly, seizures from television cartoons were parodied on animated comedy *The Simpsons*' episode "Thirty Minutes over Tokyo." Even in the twenty-first century, a tabloid reported of a baby born with insectoid "fly eyes" from the ubiquitous proliferation of video screens (Fig. 6.1).[13]

These were no longer individual problems, but ones that affected us all. Responses ranged from research, such as the Payne Fund studies of the 1930s on the relation between movies and delinquency, to public shaming, such as this cartoon satirizing the degeneracy and moral decay of radio mis-used to ask a father's blessing on his daughter's impending marriage, seeding disorder into standards of social decorum (Fig. 6.2). Such complaints are similar to those today of people sharing important news through texting.

In more philosophical theorizing, cultural critic Theodor Adorno pejoratively referred to youth as "the radio generation."[14] He described a situational

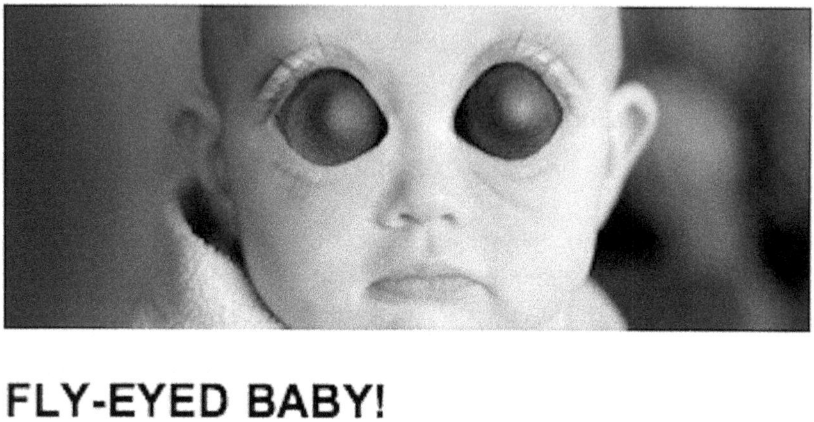

Figure 6.1. Mutant baby caused by the proliferation of video screens.

Source: Screen capture. *Weekly World News*, 2007. http://www.weeklyworldnews.com/news/breaking_news/5.

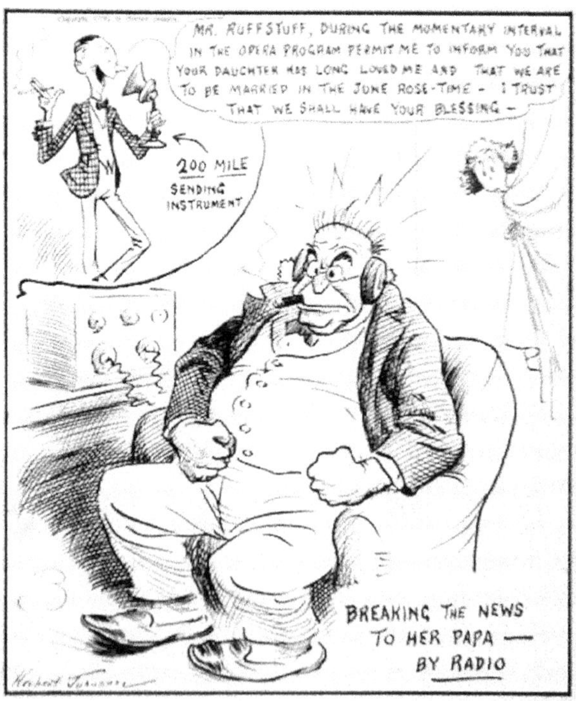

Figure 6.2. "Breaking the News to Her Papa—by RADIO." Social disorder and decay in early radio mis-use. Circa 1920s.

Source: Publication unknown. Cartoon by Herbert Johnson. Marginalia: "Future Pastimes." Library of Congress. Prints and Photographs Digital Item Display—acd1996004232-PP.png.

unawareness from the vapid and clinging melodies of popular music: "If you hum under your breath, abstracted from external things, you may run into a car at any moment."[15] Although, in this passage, Adorno's target was more on trite songs, significantly, popular music as a form depended on mass distribution through recording and playback technologies. Other areas of the academy developed to research the degenerative effects of such mass-media content as political propaganda and advertising. Although these generally were concerned with the content of communication technologies, at times they were generalized to the communication technologies overall. In this manner, they echo the technology-as-carrier model, exposing users to harmful, deleterious content. The difference being, the resulting concerns became less individual well-being and health but the cumulative, degenerative effects on society at large.

The social degeneracy of a mass-media technology has been perhaps clearest in the case of television. Its isolating, antisocial nature—compared to more social media, such as going to the movies—has been thought to induce boredom, suppress the immune system, and cause "backache, headache, insomnia, chronic fatigue, and sexual impotence."[16] Moreover, TV leads users down antisocial paths through mesmerism and hypnosis.[17] Doctors, government officials, pop culture, and other sources associate TV with dementia, cardiovascular risks, malnutrition, "video hypnosis," contributing to decreased intelligence, increased anxiety and paranoia, teen pregnancy, youth violence and inability to communicate, "feeding the egos of malcontents who go on a rampage and 'then run home to watch themselves on the 6 o'clock news'," creating a U.S. drug culture, addiction, and combining with "women's lib" and other factors so that it "develops children into passive listeners and gives them a distorted impression of the real world'."[18] A British scientist, reiterating themes of delinquency, described the use of "mind-bending sound waves, sophisticated chemicals, and television techniques to make the masses obey. This is a case of machines taking over. ... The new generation is not worried, because the youngsters have grown up with electronics and mini-gadgets."[19]

Such perspectives of situational unawareness also represent two of the most widely recognized scholars of communication theory and media studies to cross over into the popular imagination. Marshall McLuhan theorized an effect of habitual media use in which overstimulation created a "numbing" effect on the senses. This can be understood as the body's defensive reaction, blocking out the relatively new (in terms of human evolution) onslaught of mediated sensory stimuli, like squinting when walking into a bright light or covering your ears when a loud siren passes. McLuhan saw "numbing" as both the numbed audience developing an antisocial lack of empathy and also a literal, biological deadening of the senses—a very concrete technopathology.[20] Again countering the newness hypothesis, in the 1980s Neil Postman saw the forms and content of mass media, particularly television, leading to a decline in literacy, replaced by anonymous, decontextualized, dehistoricized, and uncontrollable information in a perpetual present. TV led to a degeneration of social fields requiring engagement and deliberation, such as religion and politics, and a reorganized perceptual subject who mindlessly consumed a continual flow of information, without significant engagement or comprehension.[21] Both McLuhan and Postman suggested a user beaten down, worn out, overloaded with accelerated flows of information-saturated media. They were driven to a state of catatonic shock, dull numbness, somnambulistic sleepwalking through life, in social disengagement and apathy.[22]

Perceiving sick users as degenerate social threats supports and is supported by blaming the patient. For example, in the excerpt below from a book on conspiracy theories, users are to blame for connecting to harmful elites using extra-low-frequency (ELF) mind control, in language suggesting the feminized, receptive disease. Indeed, the sufferer, given such degenerate habits here, is arguably asking for it:

> *Think about what you do that allows it in.* … Anything that opens the root chakra opens you to ELF. Drinking alcohol, smoking, illegal and legal drugs (including prescription and over-the-counter), and even some herbs open the root chakra. Anything that sexually stimulates the root chakra area *opens you up*, *ready to receive* ELF and the programming that it contains. This means anytime the genitalia responds to anything you are a candidate to receive programming, from advertisements to movies and books, to sexual activity. …
>
> The degree to which you are affected is a direct result of many factors, from your location to *your personal receptivity* to these waves. …
>
> If you are receptive to ELF, there is a part of your mind-pattern that is allowing it in. The most important part in keeping it out is correcting the mental imbalances that allow it to enter into your auric field in the first place. Like so many things, the basic mind-pattern behind allowing in ELF is a *victim mentality*.[23]

Although powerful forces purportedly have controlled all humanity for "aeons," and their mind-control technologies cause your illnesses, such sickness is nevertheless largely due to your licentious behavior, sexuality, and character defects, much like the victim of a sexual attack who dressed provocatively or, as I will discuss later, was wearing headphones.

Robert Montgomery: An Ideal Audience Member is a Healthy User

The decades during which eugenics was a widely popular social movement were also the formative decades of broadcasting. Technological innovations for telegraphy, radio, telephony, cinema, phonographs, and television rapidly accrued, with experiments in their forms and uses. Telephones were used to listen to music and news broadcasts, radio was used for one-on-one communication. Eventually each medium settled into the forms we are familiar with today and, particularly with radio, private and public struggles determined its ultimate form as a commercial, mass broadcasting medium and, in so doing, it laid the pattern followed by the later adoption and domestication of television and, arguably, the Internet.

In the case of Robert Montgomery, we see a eugenic logic at work: a concern with the self-directed evolution of audiences or, as I am conceiving of them, television users. Rather than a technophobic, academic, or other critique of the medium, he shows clearly the concern with directing the reproduction of the right sort of users—and preventing the reproduction of the wrong, degenerate sorts of users.

Robert Montgomery was a significant yet under-appreciated figure in media history. Despite his accomplishments, intriguing personal biography, and individual and familial celebrity, there has yet to be a book-length biography on the man. Yet he demonstrates a significant case of an audience member as a user. Especially with television, Montgomery positioned himself as both a producer (acting, directing, hosting, producing) and consumer (writing the book, *Open Letter from a Television Viewer*). Although the technologies of Montgomery's era were not highly interactive, I argue that his involvement in communications policy, consulting, and activism can be conceived of as interactivity. He actively engaged the medium, trying to affect aspects of its industry and content. In terms of pathological technoculture, his political consulting and public activism represent a case of eugenic concerns. He does not want people to abandon television, he wants to direct the evolution of television audiences in an idealized direction, one of healthy users.

Beginning as a New York stage actor in the 1920s, he signed a contract with MGM, one of the most powerful studios of the time, and moved to Hollywood in 1929. He worked steadily, typecast in roles as a posh playboy, appearing opposite Joan Crawford and Norma Shearer. He was nominated for a Best Actor Academy Award and New York Film Critics Circle Award in 1937 for *Night Must Fall* and in 1941 for *Here Comes Mr. Jordan*. Two years after its formation in 1933, Montgomery was elected to the first of four terms as President of the Screen Actors Guild (SAG), even contributing personal funds to support the nascent union. After serving with honors in World War II, he directed two motion pictures then, in the 1950s, moved into television, producing two dramatic series, one of which, *Robert Montgomery Presents*, he hosted. He continued to work as a producer, director, and actor into the early 1960s.

But what makes Montgomery's case more interesting is his work as an activist on behalf of film and television. Throughout, we see a eugenic concern with stopping degenerate forces in mass media. In 1947 he testified along with current SAG President Ronald Reagan as "friendly" witnesses before the House Un-American Activities Committee. SAG had originally defended actors against the accusations of HUAC, but, as it became more powerful, switched to trying to preserve the industry by aligning with HUAC. Montgomery testified that there was "a small minority in Hollywood who might be interested in Fascism or Communism." In his subsequent testimony, Reagan supported Montgomery, stating,

> My testimony must be very similar to that of … Mr. Montgomery. There has been a small group within the Screen Actors Guild which has consistently opposed the policy of the guild board and officers of the guild, as evidenced by the vote on various issues. That small clique referred to has been suspected of more or less following the tactics that we associate with the Communist Party. … Whether the [Communist] party should be outlawed, I agree with the gentlemen that preceded me that that is a matter for the Government to decide.

Like the stereotype of an inbred hillbilly family perceived as inherently prone to crime and disease, this "minority" was something to be weeded out and stopped from multiplying. Not merely because they themselves were bad, but that, through mass media, they would reproduce by infecting audiences with Communist ideas.

Exhibiting an interactive, two-way logic, Montgomery not only represented mass media to politicians, he also went on to work as a media consultant for politicians, honing their presentations to audiences. He observed Franklin Roosevelt's famous radio "fireside chats" first-hand from the White House. He coached Dwight Eisenhower during his presidential campaign and, after the election, was appointed "special consultant on TV and public communications." When Richard Nixon ran in 1960, Eisenhower offered him Montgomery's services but he declined. After his disastrous televised debate with John F. Kennedy, Eisenhower reportedly said, "(Montgomery) never would have let him look as he did in that."[24] In the 1960s, Montgomery served as communications director for John D. Rockefeller III. In these roles, Montgomery actively shaped not only media content produced by politicians but, in so doing, attempted to produce leaders who could shape the medium's policy itself. Moreover, in working to get Republican politicians elected, he can be seen as trying to use TV as a "political weapon" that can decide elections in order to direct the evolution of the citizens who comprise American audiences.[25]

In 1968 Montgomery published *Open Letter from a Television Viewer*. The book was part of a series conceived "as the platform for an international assembly of prominent people and established writers to discuss, dissect and delve into contemporary ideas and mores," such as religion, newspapers, gender, and art. Montgomery describes his "ill-tempered book" as an angry attempt to "awaken in the television viewer some conception of the dangers inherent in the free-wheeling oligopoly which dominates the mass entertainment and information given to us nearly twenty-four hours a day by network television."[26] Rather than identify a "small minority" within an industry, Montgomery now tried to change audiences, to provocatively direct their evolution into more conscious and active users of the medium. Although he claimed his open letter was addressed to the three major networks at the time, he generally referred to them in third person while directly addressing other viewers.

Much of Montgomery's discussion is technopathological. Television is described as an "air pollution … poisoning our cultural environment," one that is anti-family, undermining traditional gender roles, politically powerful, and inciting violence. Viewers were "addicts," "numb … [and] brainwashed." Montgomery described specifically the technopathology of antisocial withdrawal into media: "[The viewer] is not involved, and he is not moved to be involved by the cry for help in the streets either. So intense has his fantasy life become that he may even expect that cry to end in another commercial."[27] In a promotional article in the *National Enquirer*, he wrote that "Television is poisoning American minds. … Television must accept a large part of the responsibility for the violence in our cities today. … TV is a brainwashing machine for too many people."[28]

Exhibiting the alignment of technopathologies with other social stereotypes, Montgomery evoked delinquents when he bemoaned "juvenile domination" and linked gender and race when discussing the "ghetto of daytime television game shows and soap operas" that women watched.[29] This became even more explicit in his discussion of current events:

> The charge heard most frequently in 1968 was that television has aided and abetted the ghetto riots by carrying airborne seeds of violence across the country into every home or bar where black men could see other black men applying the torch and looting, thus inspiring others to do the same.
>
> While this is not yet a proved case, there is enough known truth in it to have made the networks exert a little caution. The riots in the summer of 1967 were given all-out coverage, while those following the assassination of Martin Luther King, Jr., were approached in a more restrained style. …
>
> Some observers of the social scene have said flatly that the power of extreme black militants is in direct proportion to the attention they get from the media.[30]

Degenerates were thought to be prone to violence, and Montgomery bemoaned that "violence is so prevalent that our times are likely to be recorded in history as the Age of Violence. If it is, I believe that network television, a medium which has preached and conveyed violence on a mass basis unparalleled in history, must take a large share of the blame." Degenerates were considered of lesser intelligence, and Montgomery blamed commercial broadcasting companies for treating their audiences "as little better than simple-minded fools." Degeneracy had a hierarchical logic, and Montgomery contrasted "the network entrepreneurs of television" with "those who work for it in the lower echelons." Although most viewers he conceived of as smarter than the fare they were served, only a small number have the intellect and initiative to call stations or write the FCC in protest. Most viewers were brainwashed by a network's self-promotional advertising and "not conscious of what mighty forces it controls."[31]

However, rather than prevent their reproduction, Montgomery's efforts can be seen as an attempt to reproduce himself in audiences as an ideal user. He urges them to become active, writing letters to officials and networks, demanding better fare. He attempts to educate them on the relationships between media, politics, policy, and government; how the public interest is undermined by the industry structure, how news distorts reality, how advertising supports the industry and corrupts the public good, and the "insanity" of the ratings system that drives the industry.[32] Reproduction of users is also seen when he expresses a particular concern for children—the first generation to come of age with the distorted reality of TV, from whom he expects ignorance. He concludes his book:

> I have reminded the reader, in every way I know how, that it is his responsibility, if he wants better television, to exercise his right as a citizen in guaranteeing that the Federal Communications Commission does the work of regulation it was created to do. … On our ability to bring these things about may well depend the political and cultural viability of this nation.[33]

He bemoans the vapidity of late-night television hosts Jack Parr, Joey Bishop, and Johnny Carson, yet goes on Carson's show to promote his book. I mention this not to suggest hypocrisy but to underscore how his crusade is about usership. There *are* good TV users. He presents himself as the example, and presumably he thought well of his own reproduction, television star Elizabeth Montgomery—who played a suburban witch on the popular sitcom *Bewitched*. Mr. Montgomery published *Open Letter* in 1968, during the very middle of *Bewitched's* 1964–1972 run on network television. The show continues to this day in syndication, with his daughter in essence taking over his creative legacy.

Stop the Sick from Reproducing: Sounding Out *Ringu+*

From midcentury production and politics, I turn now to my second example, a particular case of mass-media technopathologies that illustrate the social concern with reproduction of users as it persists in the apex of mass-media, when cinema and television are linked and expanded through home video, users are creating early forms of content, and are networked through telephones. As such, this moment illustrate the persistence of pathological technoculture into an era when mass-media broadcasting is certainly not new, and continues in popularity well into the era of the Internet.

I will examine what I refer to as "*Ringu+*," a cluster of related films and novels based on Koji Suzuki's 1991 novel, *Ringu*, The most familiar of these in the U.S. being the film *The Ring*. Whereas most scholarship on these narratives has been literary or visual,[34] I will employ a sound studies perspective to identify a central concern with reproduction of bad users. My sound studies analysis will demonstrate that, not only are the texts of *Ringu+* about usership, but specifically the reproduction of the self and others as users—bad users in particular. As suggested, this is also another refutation of the newness hypothesis: the central technologies of television and telephone are far from new at the time of these texts' creations, and, by the time of their global hit film versions, neither were VHS video cassettes. In fact, the first U.S. film version was released in October of 2002, only five months before weekly video rental income for VHS tapes would be surpassed, for the first time, by DVDs, the fastest-adopted consumer electronics item at that time.[35]

Although the original three novels and story collection—in English, *Ring*, *Spiral*, *Loop*, and *Birthday*—do not unfold in a linear chronology, I will summarize their events in temporal order for clarity.[36]

Scientists working on artificial intelligence create an electronic world called The Loop. Although it mirrors our world, it is not a virtual world in the sense of a fake environment with which we engage, but more an electronic Petri dish or terrarium in which sentient digital lifeforms are born and evolve in an accelerated but parallel path that mirrors human evolution. The Loop is a digital reflection of humanity, yet wholly autonomous. At one point in Loop history, a psychic woman is publicly humiliated and commits suicide. Her bastard child, Sadako, who is intersexed and has even stronger psychic powers, attempts a career as a stage actress but also suffers humiliations, a sexual assault, and a fatal attempt on her life. Furious with the world that has hurt her, and with her own inability to have children, just before her death she mentally wills her rage into an open television-signal frequency and onto a videotape being recorded. Burned into the filings of the magnetic tape are a disturbing series of images from her life. This tape infects its viewers with a "ring-virus" that kills them in a week if they do not make a copy of the tape and show it to some new user, reproducing the cycle.

As the tape-viewing spreads virally, Sadako's spirit manages to impregnate a woman and enter the world of the living. Reporters investigating the fatal videotapes write up the story, which becomes a hit book and movie. However, these mediated versions of the narrative also carry the same viral curse as the videotapes. Through their spread, and the methods by which Sadako has learned to physically reproduce herself, Sadako clones repopulate and eventually take over the world, snuffing out genetic diversity and killing the human race.

However, all of this is only within the electronic world of The Loop. Scientists observe its events through computer visualizations, watching the virtual world unfold at its accelerated pace, and revisiting its past. (Although the Loop world has died, because it was all digital data, its every moment and inhabitant were stored, so Loop time can be rewound and replayed.) Unfortunately, at one point the ring-virus crosses over from the Loop into the "real" world, unleashing a cancer-like plague that threatens to destroy our world. Through communications and character-transfers between the two worlds, interventions in the events of Sadako and the ring-virus halt their progress, restoring balance to the Loop world and stopping the cancer plague in ours.

Numerous films have been made based upon these books, such as *Ringu*, *Rasen*, *Ringu 2*, *Ringu 0: Batsuday*, *Sadako 3D*, *Sadako 3D 2*, and *Sadako vs. Kayako* in Japan; *The Ring Virus* in South Korea, and *The Ring*, *The Ring Two*, *Rings* (an interstitial short film) and *Rings* (a feature) in the US. In Japan, there have also been television versions; throughout Asia and in the West there have been many films, TV episodes, and anthologies exploring very similar stories and themes, such as *Chain* (two Japanese films), *One Missed Call* (three Japanese films, a television series, and one U.S. film), the *Pulse* films (one Japanese and three US), *Phone*, *Txt*, *Text*, and *Dial D for Demon*. The first *Ringu* film was unprecedented both in its local popularity and international impact, sparking not only these imitators, but also stylistic innovations in the horror genre and a new uptick globally in the cyclical popularity of the horror genre. In Japan, it even spawned its own health panic, with Sadako becoming a common vision in a wave of *kanashibari* or childhood night terrors.[37]

More an interdisciplinary conversation than strict method, sound studies does not displace visual, textual, or other types of analysis, but elevates the inclusion of sonic artifacts, practices, and experiences as both objects of study and critical modes. It augments with sound. The primary element of a sound studies approach is, simply, paying attention to sound. Looking for and listening to sound helps one attune to significant moments in the narratives and major fulcrums of the plots in *Ringu+* that might not be noticed or noticed in the same way if one simply takes them as visual narratives, which their orientations around looking at videotapes and a virtual world might suggest. Yet, sound studies readies one to notice representations of sound at key moments: The title's English homophone of multiple meanings across the same sound—"ring" as geometric circle, cycle of repeated events, wedding band, or hailing sound of a telephone—is not an accident of translation but explicitly noted in the book, and each of these meanings are relevant. In the first cases of ring-virus death, a warning of impending death is announced by a mystery phone call, moments after viewing the tape. The appearance and disappearance of these calls are a crucial clue in geographically uncovering the mystery. The psychic powers of hero-monster Sadako and her mother, which attract to them their humiliations and sufferings that incite the narratives, are indicated by a high, metallic screech or whine, a sound that recurs in the imitation films as death rattles and gasps of telephonic spirits and demons. Sadako's public humiliation, and the ruining of what appears to have been her one chance at happiness, has to do with the inappropriate public playing of audio tapes, implicating betrayal by her one and only potential lover, a theater troupe's sound effects man. Moreover, the telling of this backstory serves in the books and films as what humanizes Sadako, transforming her from a wrathful spirit into a more sympathetic monster, in the classic tradition of gothic horror, such as *Frankenstein*. Sound, therefore, is not only a signature marker of the series and its villain/heroine, it is central to the inciting events of the narrative, and those are the narratives by which sympathetic intersubjective understanding of the villain, and thereby her redemption, are achieved. Significantly, the final image of the books is one of intersubjective communion bereft of visual contact. Furthermore, the telephone rises in significance from an adjunct technology in the first *Ring* narratives to the central technology in many *Ring*-inspired films, such as *Phone*, the *One Missed Call* films and television series, *Txt*, and *Dial "D" for Demon*, suggesting the impact of the aural aspects of *The Ring* narratives.

The primary theme of *Ringu+* is that communication entails contagion. The conduits of human connectivity are also the carriers of inhuman carrion. Instead of an uncivilized Western savage microbe, the Other is the otherworldly realm of the dead.

In addition to elevating attention to representations of sound, sound studies also theorizes the work of sonic practices. This provides a theoretical tool to further examine the prominence of sound in *Ringu+*. In specific, I draw on a large body of scholarship in which sound and aural practices have been theorized as constitutive of subjects and subjectivities. For example, Jon Cruz has analyzed slave songs as a crucial empathetic link in the abolitionist movement. Their sounds and vocalizations not only provided for expression, community, and identity building among American slaves, it also, he argues, was a major medium by which whites came to see them as fully human. It communicated a degree their subjectivities, reproducing a degree of their suffering for whites to experience.[38] Cruz' work is representative of a body of sound studies work theorizing constitution of subjectivities, which aids in reading *Ringu+*. Aural practices such as improvisation, spoken word, and verbal play have been examined as processes of collective identity formation.[39] Indeed, race has been a heavily analyzed vector of sonic subject formation.[40] Gendered, sexed, and sexual subjectivities have also been theorized in feminized practices of consuming audio, fan practices, and even in the very structure of Western serial composition.[41] The sonic formation of broader subjective categories, such as past/present, real/fantastic, geographic/personal, and Self/Other, is explored in, for example, Josh Kun's work around Tijuana[42] and others' analyses of exotica music.[43]

What subjects, then, are being produced in *Ringu+* narratives? Technological subjects. Users. *The Ring Virus*, the Korean version of the first novel and the film that hews most closely to the books, opens with the camera slowly tracking across a variety of communication technologies. From a close-up of analog television static (similar to the iconic opening shot of *Poltergeist*), the camera creeps upstairs to a bedroom, showing a VCR, CDs, computer, cell phone, pager, and personal stereo. The tracking shot terminates with a young girl unnaturally contorted atop her bed. Although we soon realize she is doing yoga, her awkward position foreshadows the fatal convulsions the virus causes, as well as the possession contortions in much *Ringu*-inspired Asian horror, such as the *One Missed Call* films and moments in the *Ju-On* series. Moreover, her location on a bed suggests sexual positions, and, as I will describe, reproduction is a central theme of *Ringu+*. The girl's unnatural state is linked with media use from the establishing tracking shot and a pager that commences buzzing the moment we see her in this position. This underscores the film as a narrative of usership, as does the pager's message, which we will learn later to have been, "Did we see what we were not supposed to see?" After this shot, a friend calls and, during the course of their conversation, dies from the virus. She listens to his death throes as her TV turns itself back on, and a white-out signals her death from the virus.

Moreover, some of these users are, quite pointedly, sick. The users of *Ringu+* are literally ill, physically and mentally. They watch a tape, receive a call, or read a book and become nauseous, sickened, anxious, and emotionally labile, and then often die. The screech of Sadako and her mother is painful, like extreme tinnitus: In a scene from *Ringu 2*, patients at a mental hospital become agitated and violent from hearing it.

In various *Ringu+* plots, characters are hospitalized or institutionalized from traumas associated with aspects of the ring virus. Moreover, the books reveal that this is not psychosomatic or a purely ethereal, psychic curse: the ring virus is a real virus, a hybrid of Sadako's DNA and the smallpox virus that was reproduced in her from her rapist, which then becomes a form of cancer in the "real" world. Indeed, the disease of smallpox is, like sound, crucial to the events of *Ringu+*. Sadako's death and rape occur on the site of a former smallpox colony, her rapist is a former doctor there and the last surviving smallpox-infected person in Japan. It is also where the first viewing of the infected tape occurs.

Moreover, it is on this site where bad users of technology start the plague: In the novels (unlike the films) there is a patient-zero videotape, an epidemiological epicenter. The tape, when first viewed, had instructions at the end explaining that, to save themselves, they must copy the tape and show it to someone else. The vacationing teenagers who first view Sadako's creation, thinking it a joke, intentionally erased over the instructional section of the tape, imagining that this would make it even scarier to those who found and viewed it next. The Patients Zero of the ring-virus epidemic are technology users who *do not follow the directions*—bad users—and apocalyptic horrors result. Indeed, users are given directions throughout *Ringu+*: "You should screen your calls," schoolgirls warn in the Japanese *One Missed Call*. "Don't look at the phone," warns a woman blinded by the curse on her *Phone*. Direction, protocol, and procedures multiply as teens tell each other how the ring curse works, what you're supposed to do and not do, and provide instructions for participating in *Rings*. Not doing what you're supposed to do, as the horrified son tells his mother who has just freed the American version of Sadako in *The Ring*, puts everyone at risk. Indeed, one literary scholar has described *Ringu* as "a moral apocalypse" and "a parable about social responsibility."[44]

Ringu+ narratives are obsessed with reproduction. The Loop is a reproduction of our world. Sadako is the offspring of a psychic woman and a sick man, reproducing their traits. Although Sadako's pathologization evokes earlier discussions of feminine neediness and sensitivity—her desire for love evoking Louise from *Possessed* or her daughters, the crazy jezebels and clinging vines of *Poison Ivy*, *Fatal Attraction*, *Hand that Rocks the Cradle*, and *Obsessed*—the true terror of Sadako lies in her reproductive capabilities, which challenge heteronormative patriarchy through various forms of asexual reproduction. Sadako reproduces herself as a pregnancy in one woman, as a virus on the tape, and the tape reproduces itself through news stories and films. Reincarnating a central character's dead son is the mode by which a human becomes a traitor to Sadako's cause; another character is reproduced both within the

Loop world and in our world. Another Loop character is reincarnated in our world as a scientist's son. The "cure" for persons infected with the ring virus is to reproduce the tape and an audience for it. Suzuki notes in his biographical statement that he is "a respected authority on childrearing and has written numerous works on the subject" and wrote *Ring* "with a baby on his lap." Reproductions continue in the films, themselves reproductions of the novels, which also prominently feature reproduction of the curse and its secrets through phone calls, gossip, and web-based clubs of thrill-seeking teens.

Finally, *any* narratives about communication media evoke reproduction. As mentioned previously, broadcasting's semantic roots lie in agriculture—reproducing crops—and evangelizing—reproducing the word of God and his subjects.[45] As Walter Benjamin famously explored, communication media are inherently about reproduction of formerly singular cultural objects.[46] Furthermore, each "new" communication medium reproduces certain formal qualities and usage practices of previous media.[47]

Feminist scholar Angela McRobbie has argued that moral panics around sexuality often are not so much about sex in and of itself, but about reproduction and childbirth as markers of being lower class, about motherhood without sufficient support so that cultural and economic productivity is hindered.[48] One does not have to look far for anecdotal evidence of this argument, such as the mass-to-porn media careers of *Teen Mom* star Farrah Abraham and "Octomom" Nadya Suleman. In the latter case, her status as a "welfare mom" fueled the transformation from multiple-birth starlet to object of scorn, with copies of her welfare card reproduced on gossip blogs. On the flipside, Amy Chua's *Battle Hymn of the Tiger Mother* drew controversy for espousing authoritarian, achievement-oriented mothering.

The *Ringu+* narratives condemn bad users, not merely who technologically connect with supernatural forces, but who also cannot control the reproductive *consequences* of connecting. The problem is not simply the bad girl, but the bad, pregnant girl who produces more bad subjects, draining the welfare state, swelling the lower classes, being unproductive, ending diversity and life in the Loop, and infecting the "normal" population. Robert Payne has used the term "promiscuous" to describe network culture, in the sense of a surplus of injudicious connects. However, I would extend the concept of promiscuity to include the sexual meanings of the word as well. The promiscuous user connects excessively and without good judgement, exceeding boundaries that were put there for all of our safety, bringing demons and cancers into the realms of those who were good to begin with. Yet she is also reproductive. She is Patient Zero, the deserving sufferer whose promiscuity passes her disease on to "innocent victims," as with crack babies, children online, and heterosexuals.

It's not just about hooking up to bad things that hurt you, it's about *reproducing more bad things*: unproductive subjects, cycles of sexual abuse, myths of homosexual recruitment, compulsions and addictions. Explicitly evoking degenerate substance abusers, addiction to or dependency upon electric communication technologies is a frequent topic of news and entertainment.[49] Bad reproduction is that which produces unproductive subjects. Addicts reproduce addicts by example and by luring others into using, gaining access to more resources and substances and ameliorating their own guilt and anxieties by increasing their circle of fellow users, all of which leave productive economies for unproductive ones, eventually. The pregnant mother is removed—for as little time as possible—from the labor market. Reproduction hurts productivity and consumption. Unauthorized reproductions online threaten the stability and integrity of consumers' market profiles. Do we target-market to the sexting girl with ads for Barbie, Victoria's Secret, or *Seventeen* magazine? Moreover, her inappropriate technological use risks the greatest unproductivity at all: her disconnection from the network, her removal from the online economy, whether by parental punishment or legal retribution. Discourses of bad connections sidestep concerns and controversies regarding sex education, childcare, birth control, and access to health care, that contribute to unproductive reproduction.

In *Ringu+*, the ultimate problem involved is reproduction without control, a needy woman whose desire to reproduce is unproductive to the point of destroying two worlds. This gendered logic applies to technology. The healthy user is connected, but knows her limits. These limits apply to not only with whom she hooks up, but the resultant reproductions. Reproduction in this case being, not in the sense of children, but in terms of reproductions of herself and subjectivity. As media scholar Wendy Chun writes,

> Rather than simply allowing people to exercise what Walter Benjamin once called their "legitimate claim to be reproduced," the Internet circulates their "reproductions" without their consent or knowledge. … Computation's rampant reproductions … literalize control. This control gives users greater access to each other's reproductions. Putting [internet traffic reading software] sniffers into "promiscuous mode," for instance, accesses all the traffic going through a cable.[50]

Technopathologies indicate the impossibility of a human "promiscuous mode"—we cannot monitor all traffic, all circulation. We know our unauthorized reproductions are in controllable, uncontrollable, and ambiguous circulation, from our gaming avatars visiting the homes of friends to chain-postings warning of our personal photos appearing in Facebook ads. Unsurprisingly, at least one technopathology involves blaming TV for contributing to teen pregnancy.[51]

Conclusion: Broadcasting and Networking

Broadcasting is not archaic, old media. It has not been displaced. As the transitional moment of *Ringu+* suggests, it has been integrated into the contemporary communications network, along with its technopathologies. In *The Signal*, an unknown electrical force spreads throughout all communications systems, infecting users with paranoid, homicidal rage. Addiction, hallucinations, and possession are due to spirit-infused television signals in the film *Dead Waves*. Their broadcasts possess viewers with depression, hostility, and suicide. While authorities debate whether they are just mentally ill, possessed viewers infected others with involuntary exposure, forcibly holding their eyes open.

Networking increases the threat of technologies because multiple technologies are connected to each other and to users. These can carry deleterious threats or cause injury themselves. The network is contagious, and all on it are at risk. However, the greatest risk may not be exposure to dangers, but the reproduction and proliferation of those dangers. The global communications network is not merely a massive series of conduits for infection, it is also a culture for growing them. The electronic global village is a Petri dish, in which bad users do not merely sicken, they also reproduce.

With the rise of the Internet, mass media communications technologies have evolved into components of a larger computerized network, and technopathological discourse persists in many components of our current media ecosystem, such as computers, game systems, and the networked whole. For example, both television and computer screens have been associated with insomnia.[52] More ominously, a *National Enquirer* front page bellowed, "COMPUTERS ARE TURNING US INTO A NATION OF LAME BRAINS." In the story, a retired Air Force Senior Management Analyst of 15 years warned:

> Unless we are careful, the computer may soon warp our thinking abilities as a nation—just as the automobile has withered out walking ability. ... The growing trend today of using computers to make important decisions in government, business,

> and industry is one of the deadliest illnesses of our society. … Computers are like electronic Frankenstein monsters.[53]

Similarly, John G. Kemeny's *Man and the Computer* warned that humanity was evolving into a new species in which computers are our symbionts, and a tabloid warned that "Computers Are Turning Kids Into Dropouts."[54]

Electronic games are mass-produced, yet individually consumed, and with social scoring at least or interactive partners. They have provoked techno-pathological discourse across their varied forms, often linked with television and other mass media. Instructions for games such as Sega Genesis' *Sonic the Hedgehog 2*, Sega's *Mickey Mania* and Nintendo *Wii Sports* warned of repetitive motion injuries, eyestrain, seizures, motion sickness, and other physical reactions. Addiction to video games appeared in the television series, *Square Pegs*, and the films *Nightmares*, *Maximum Surge Movie*, and *Hellraiser: Hellworld.* Celebrity rehabilitation facility Promises Treatment Center in Malibu has offered treatment for treating video game addicts.[55] Dr. Philip Nader, a University of California-San Diego pediatrics professor emeritus conducted "one of the largest studies of its kind" on obesity, partially attributing decreasing physical activity to the increasing popularity of computer and video games.[56] A University of Buffalo study showed that restricting children's video game and television time also decreased their individual Body Mass.[57] A tabloid reported similar studies, warning, "TV Makes Kids Fat!"[58] and the "Make Love not Warcraft" episode of *South Park* depicted an addictive Massively Multiplayer Online Role-Playing Game (MMORPG) causing obesity and acne.[59] An Assistant Professor of Pediatrics warned that a video game based on a horror movie "'legitimizes cruel and inhuman behavior. … It could feed the fantasies of seething idiots and degenerates'."[60] Indeed, a tabloid report claimed that "Pac-Man Fever Drives Teenager to Burn His House Down."[61] In the film *Brainscan*, a video game draws users into a "mind entry program" a state "not unlike hypnosis" through a pulse "embedded in the TV's blanking signal." Flashing lights seem to hypnotize user, who commit murders in the real world under the spell of Trickster, a supernatural game master. In some fictional narratives, the deaths of characters in video games even cross over to deaths of their players, as in the film *Stay Alive*.

Finally, all of these components come together in today's networked media environment. For example, conspiracy subcultures abound with stories of mass communication technologies and related devices misused by visible and shadow governments, extraterrestrials, and others to make people suffer or succumb to mind control.[62] A conspiracy author described how extra low frequency (ELF) electromagnetic bombardment from satellites, TV, radio, computers, and other communication technologies was used by the 13 families of the Illuminati to subjugate humanity, inducing mind-numbing, tinnitus, depression, aggression, fear, paranoia, dissociation, multiple personalities, changes in behavior, moles and bodily markings, changed vision, blank expression or affect, nightmares, obsessive-compulsive behavior, substance abuse, and amnesia.[63] Such narratives appear in mainstream popular culture as well. The Arcangel meta-network of satellite and mobile communication technologies is used to hypnotize the masses in "The Sound of Drums" episode of *Dr. Who*. Omniscient, homicidal surveillance technologies of government conspiracy appear in the film *Eagle Eye*. Hypnotic mind control and fatally sickening computer technologies developed by TV advertising companies appear in the film *Looker*. In the film *Halloween III: Season of the Witch*, evil druids embed a signal in TV commercials that trigger fatal devices within Halloween masks to kill the wearer. Addiction appears with hallucinations and physical mutations due to a hidden television signal designed by elite techno-vigilantes to purge society of porn-watching deviates in the film *Videodrome's* cable broadcast. One of the many threads of the conspiracy series *The X-files* involved, in the episode "Wetwired," an electrical signal embedded in a cable television broadcast and videotapes of it inducing violent psychosis, hallucinations, anxiety and dementia. The song "World Destruction" by Time Zone, a collaboration between hip-hop pioneer Afrikkaa Bambataa and Sex Pistol John Lydon, listed technopathological conspiracies among manifold cultural problems including racism, religious fundamentalism, and nuclear war: "The CIA is looking for you. / The KGB is smarter than you think. / Brainwash mentalities to control the system. / Using TV and movies—religions of course." In 2007, Richard Hawley's "Tonight the Streets Are Ours," warned that "They make our TVs blind us / From our vision and our goals."

However, as the clichéd joke goes, "Just because I'm paranoid doesn't mean they're not out to get me." A discourse being "paranoid" should not imply a lack of impact and effects. One man in my focus groups thought Internet users should self-monitor and restrict where they went online: "If they go on the Internet to the wrong place, they can get contacted by the evil FBI or CIA, if they're going through the wrong places."

Such technopathologies continue the discourses of usership seen in Robert Montgomery and *Ringu+*. In networked, mass, interactive media, sick users threaten to spread sickness and reproduce more unhealthy users.

Notes

1. Foucault, "Truth and Power," 125.
2. Kennedy, *History*; Gilman, *Difference and Pathology*; Martindale, "Degeneration"; Walter, "Degenerate?"
3. Foucault, "19 March 1976," 315; "Dangerous Individual."
4. Martindale, "Degeneration," 180.
5. Standage, *The Victorian Internet*, 140, 132, 13.
6. Foucault, "19 March 1976," 316.
7. Ibid. 317.
8. Ibid. 317–318.
9. Hilmes, *Radio Voices*.
10. Browning, *Infectious Rhythm*.
11. Weiss, *Phantasmic Radio*.
12. Kerbel, "3-D or Not."
13. Stewart, "Fly-eyed Baby!"
14. Adorno, *Essays on Music*, 462.
15. Ibid. 506.
16. Haley, "'Dallas'."
17. Bannister, "You Can be Hypnotized"; Brecher and Brecher, "Too Much TV."
18. Aronson, "Excessive Television," Balfour, "Professor"; Bannister, "You Can be Hypnotized," Harrison, "Television"; "Hayakawa Blames TV for Riots and Looting"; *Manhattan*; Mahoney, "Psychoanalyst"; Nelson, "TV Creates"; Ramierez, "Warning," Sidney, Sternfeld, Haskell, Jacobs, Chesney, and Hulley, "Television Viewing"; *The Tube;* "Too Much TV Watching," "What's in the Box?"
19. Calder, "World-Famous," 24.
20. McLuhan, *Understanding*, *Essential*.
21. Postman, *Disappearance*.
22. The biological symptoms of this are similar in many ways to neurasthenia.
23. Swerdlow, *13-Cubed*, 320, 330, 332; all emphases mine.
24. Lee, *Earl*, "The People You Meet" section, ¶4.
25. Montgomery, *Open Letter*, 60, 72.
26. Ibid. 7, 11.
27. Ibid. 151, 35, 29.
28. Terrence, "Television."
29. Ibid. 40.
30. Ibid. 113–114.
31. Ibid. 12, 14, 26.
32. Ibid. 136.
33. Ibid. 152.
34. Wee, "Visual Aesthetics".
35. "It's Unreel."
36. Suzuki has since published new books about Sadako, but these and their related films deviate considerably from the core narratives I examine here, focusing more on ghosts than the technologies that carry them.

37. Schegoleva, "Sleepless in Japan."
38. Cruz, *Culture on the Margins*.
39. Becker, "Etiquette"; Keil, "Motion and Feeling"; Mackey, *Bedouin*, "Discrepant."
40. Gilroy, "Between the Blues"; Gray, *Cultural Moves*; Henriques, "Sonic Dominance"; Lipsitz, *Dangerous Crossroads*; Moten, *In the Break*; Rose, *Black Noise*; Weheliye, "'Feenin'," *Phonographies*.
41. Adorno, *Essays on Music*; Gottlieb and Wald, "Smells Like"; Kostenbaum, *Queen's Throat*; McClary, *Feminine Endings*, "Narrative Agendas": Wald, "'I Want'."
42. Kun, *Audiotopia*.
43. Lanza, *Elevator Music*; Taylor, *Strange Sounds*.
44. Jones, "Video Nasty," 215.
45. Hamilton, "Unearthing Broadcasting."
46. Benjamin, "Work of Art." Although, as Weheliye (*Phonographies*) notes regarding Benjamin, the original is perhaps weakened by the copy, but the very notion of the original's aura is also constituted by and requires the copies, making the reproduction of copy/original actually a two-way, mutual re-production.
47. Grusin and Bolter, *Understanding*; Manovich, *Language*; McLuhan, *Essential*, *Understanding*.
48. McRobbie, *Aftermath*.
49. Popular-culture examples include addiction to social media gaming in the film *#Horror*, to video games and the internet leading to child neglect in the *Law & Order: SVU* television series episode "Bullseye," and to social media in the "Be Right Back" episode of the UK series *Black Mirror*. News articles have discussed addiction in terms of BlackBerry personal digital assistants aka "Crackberries" (Dvorak, 2014), college students and texting (Rice, 2011), and television producing anxiety (Brecher and Brecher, 1987). One tabloid quoted a New York psychiatrist that, in 1996, 150,000 people were already addicted to the Internet, "costing people their jobs, homes & families" (Smith, P., 1996, headline).
50. Chun, *Control*, 4.
51. Harrison, "Television."
52. Horse, "Insomnia," 34.
53. Cassidy, "Psychiatrist," 5.
54. O'Neill, "Computers."
55. Bargar, Smith, and Wilson, "Gamers Anonymous."
56. "Restricting Kids' Video Time."
57. Shrieves, "Phys-ed Flop."
58. Fitz, "TV Makes Kids Fat."
59. Some scholars suggest that this recent obesity epidemic is another cultural complex, in which social issues, such as proper citizenship and national security, are displaced onto concerns with improper embodiment (Shugart, "Consuming Citizen"). While important to consider, solely examining these technopathologies as panics and social displacements serves to pathologize the users who suffer from them, and obscures cultural work of technopathologies.
60. Regan, "Game of Shame," 61.
61. Meyers, "Pac-Man."
62. Dean, *Aliens*; Rohrlich, "They Say Government," Swerdlow, *13-Cubed*; Tuckett, *Conspiracy Theories*; X, *Mind Stalkers*.
63. Swerdlow, *13 Cubed*.

References

#Horror. Dir. Tara Subkoff. 2015, USA.

Adorno, Theodor. *Essays on Music*, University of California Press, 2002.

Aronson, M. "Does Excessive Television Viewing Contribute to the Development of Dementia?" *Medical Hypotheses*, no. 41, 1993, pp. 465–466.

Balfour, M. "Professor of Psychiatry Advises … Turn off TV and Ignore Experts to Raise Well-adjusted Children." *National Enquirer*, 1973, October 14, p. 8.

Bannister, P. "You Can Be Hypnotized by TV." *National Enquirer*, 1981, November 10, p. 34.

Bargar, Lesley, R. J. Smith, and Sara Wilson. "Gamers Anonymous." *Los Angeles Magazine*, 2008, p. 34.

"Be Right Back." Dir. Owen Harris. Episode of *Black Mirror*. 2013, UK.

Becker, Howard S. "The Etiquette of Improvisation." *Mind, Culture, and Activity*, vol. 7, no. 3, 2000, pp. 171–176.

Benjamin, Water. "The Work of Art in the Age of Mechanical Reproduction." In *Visual Culture: The Reader*, edited by J. Evans and Stuart Hall, Sage, 2005.

Brecher, H., and A. Brecher. "Too Much TV can Trigger … Anxiety." *National Enquirer*, 1987, December 8, p. 24.

Browning, B. *Infectious Rhythm: Metaphors of Contagion and the Spread of African Culture*, Routledge, 1998.

"Bullseye." Dir. Peter Leo. Episode of *Law and Order: Special Victims Unit*. 2010, USA.

Calder, I. "World-Famous Scientist Warns: Govt. Has New Techniques that Can Control Your Mind." *National Enquirer*, 1969, May 25, p. 24.

Cassidy, J. "Psychiatrist Says Long-Distance Air Travel Causes World Leaders to Make Dangerous Decisions." *National Enquirer*, 1969, November 24, p. 2.

Chun, Wendy Hui Kyong. *Control and Freedom: Power and Paranoia in the Age of Fiber Optics*, MIT Press, 2006.

Cruz, Jon. *Culture on the Margins: The Black Spiritual and the Rise of American Cultural Interpretation*, Princeton University Press, 1999.

Dean, Jodi. *Aliens in America: Conspiracy Cultures from Outerspace to Cyberspace*, Cornell University Press, 1998.

Dvorak, John C. "End of the Road for Research in Motion." *Marketwatch*, 2014, September 22. http://www.marketwatch.com/story/end-of-the-road-for-research-in-motion-2011-09-16. Accessed September 16, 2011.

Fitz, R. "TV Makes Kids Fat." *National Enquirer*, 2007, March 22, p. 46.

Foucault, Michel. "About the Concept of the 'Dangerous Individual' in 19th-Century Legal Psychiatry." *International Journal of Law and Psychiatry*, vol. 1, 1978, pp. 1–18.

Foucault, Michel. "19 March 1976." In *"Society Must be Defended": Lectures at the Collége de France, 1975–76*, edited by Francois Ewald, M. Bertani, and A. Fontana, Picador, 2003, pp. 291–321.

Foucault, Michel. "Truth and Power: An Interview with Alessandro Fontano and Pasquale Pasquino." In *Michel Foucault: Power/Truth/Strategy*, edited by Meaghan Morris and Paul Patton, Feral Publications, 1979, pp. 48–58.

Gilman, Sander L. *Difference and Pathology: Stereotypes of Sexuality, Race, and Madness*, Cornell University Press, 1985.
Gilroy, Paul. "Between the Blues and the Blues Dance: Some Soundscapes of the Black Atlantic." In *The Auditory Culture Reader*, edited by Michael Bull and L. Back, Berg, 2003.
Gottlieb, J., and Gayle Wald. "Smells Like Teen Spirit: Riot Grrrls, Revolution and Women in Independent Rock." In *Microphone Fiends: Youth Music and Youth Culture*, edited by Andrew Ross and T. Rose, Routledge, 1994.
Gray, Herman. *Cultural Moves: African Americans and the Politics of Representation*, University of California Press, 2005.
Grusin, Richard, and Jay David Bolter. *Remediation: Understanding New Media*, The MIT Press, 1999.
Haley, L. "'Dallas' and 'Dynasty' Can Really Make You Sick." *National Enquirer*, 1987, March 17, p. 27.
Hamilton, J. "Unearthing Broadcasting in the Anglophone World." In *Residual Media*, edited by Charles R. Acland, University of Minnesota Press, 2007.
Harrison, L. "Television is Contributing to Teen Pregnancies." *National Enquirer*, 1980, August 12, p. 32.
"Hayakawa Blames TV for Riots and Looting." *National Enquirer*, 1969, May 4, p. 29.
Henriques, J. "Sonic Dominance and the Reggae Sound System Session." In *The Auditory Culture Reader*, edited by Michael Bull and Les Back, Berg, 2003.
Hilmes, Michelle. *Radio Voices: American Broadcasting, 1922–1952*, University of Minnesota Press, 1997.
Horse, Lester the Talking. "Insomnia the Cause?" *Weekly World News*, 2007, July 30, p. 34.
"It's Unreel: DVD Rentals Overtake Videocassettes." *The Washington Post*. 2003.
Jones, Steve. 2012. "Video Nasty: The Moral Apocalypse in Koji Suzuki's *Ring*." *Lit: Literature Interpretation Theory*, vol. 23, no. 3, pp. 212–225.
Keil, C. "Motion and Feeling Through Music." In *Music Grooves*, edited by C. Keil and S. Field, University of Chicago Press, 1994.
Kennedy, Michael. *A Brief History of Disease, Science and Medicine from the Ice Age to the Genome Project*, Asklepiad Press, 2004.
Kerbel, M. "3-D or Not 3-D." *Film Comment*, vol. 6, no. 6, 1980, pp. 11–20, 80.
Kostenbaum, Wayne. *The Queen's Throat: Opera, Homosexuality and the Mystery of Desire*, Poseidon Press, 1993.
Kun, Josh. *Audiotopia: Music, Race, and America*, University of California Press, 2005.
Lanza, Joseph. *Elevator Music: A Surreal History of Muzak®, Easy-listening and Other Moodsong®*, revised and expanded edition, University of Michigan Press, 2004.
Lee, R. E. *The Earl of Hollywood: The Life and Career of Robert Montgomery*. 2004. http://www.earlofhollywood.com. Accessed June 6, 2013.
Lipsitz, George. *Dangerous Crossroads: Popular Music, Postmodernism and the Poetics of Place*, Verso, 1994.
Mackey, N. *Bedouin Hornbook*, University of Kentucky, 1986.
Mackey, N. *Discrepant Engagements: Dissonance, Cross-culturality, and Experimental Writing*, Cambridge University Press, 1993.

Mahoney, R. "Psychoanalyst Blames Unreal World of TV for Today's Violent Youth." *National Enquirer*, 1969, April 2.

Manhattan. Dir. Woody Allen. 1979, USA.

Manovich, Lev. *The Language of New Media*, The MIT Press, 2001.

Martindale, Colin. "Degeneration, Disinhibition, and Genius". *Journal of the History of Behavioral Sciences*, vol. 7, no. 2, 1971, pp. 177–182.

McClary, Susan. *Feminine Endings: Music, Gender, and Sexuality*, University of Minnesota Press, 1991.

McClary, Susan. "Narrative Agendas in 'Absolute' Music: Identity and Difference in Brahms's Third Symphony." In *Musicology and Difference: Gender and Sexuality in Music Scholarship*, edited by R. A. Solie, University of California Press, 1993.

McLuhan, Marshall. *Essential McLuhan*, Basic Books, 1995.

McLuhan, Marshall. *Understanding Media: The Extensions of Man*, The MIT Press, 1994.

McRobbie, Angela. *The Aftermath of Feminism: Gender, Culture and Social Change*, Sage, 2009.

Meyers, J. "Pac-Man Fever Drives Teenager to Burn His House Down." *National Enquirer*, 1983, March 15, p. 7.

Montgomery, Robert. *Open Letter from a Television Viewer*, J. H. Heinemann, 1968.

Moten, Fred. *In the Break: The Aesthetics of the Black Radical Tradition*, University of Minnesota Press, 2003.

Nelson, G. "TV Creates a Drug Culture in U.S." *National Enquirer*, 1972, January 2, p. 2.

O'Neill, B. "Computers are Turning Kids into Dropouts." *National Enquirer*, 1980, July 23, p. 18.

Postman, Neil. *The Disappearance of Childhood*, Delacorte Press, 1982.

Ramierez, Thomas P. "Warning—Too Much TV Can Trigger Fear & Stress." *National Enquirer*, 1987, July 27, p. 21.

Regan, J. "Games of Shame: Sick Toys Tell Kids Drugs, Murder and Cheating are OK." *National Enquirer*, 1983, January 18, p. 61.

"Restricting Kids' Video Time Reduces Obesity, Randomized Trial Shows." 2008. *Science Daily*. http://www.sciencedaily.com/releases/2008/03/080303164504.htm. Accessed March 4, 2008.

Rice, Alexandra. "Bleary-Eyed Students Can't Stop Texting, Even to Sleep, a Researcher Finds." *The Chronicle of Higher Education*. 2011, November 20. http://chronicle.com/article/Tired-Students-Skip-Sleep-for/129838/. Accessed June 15, 2016.

Rohrlich, Justin. "They Say Government Mind Control Is Real—And That They're Part of It." *The Daily Beast*. 2017, January 30. http://www.thedailybeast.com/articles/2016/09/04/they-say-government-mind-control-is-real-and-that-they-re-part-of-it. Accessed July 4, 2017.

Rose, Tricia. *Black Noise: Rap Music and Black Culture in Contemporary America*, Wesleyan University Press, 1994.

Schegoleva, A. "Sleepless in Japan: The Kanashibari Phenomenon." Paper presented at the Kenkyu Postgraduate Research Seminar in Japanese Studies Oxford Brookes University Research Centre, Oxford Brookes University, UK. 2001, July 28.

Shrieves, L. "Phys-ed Flop: Study Confirms Teens' Dramatic Couch-potato Trend. Raising Couch Potatoes: Teens Veg Out." *Orlando Sentinel*, 2008. http://www.orlandosentinel.com/orl-activekids08jul16,0,7578989.story. Accessed July 16, 2008.

Shugart, Helene A. "Consuming Citizen: Neoliberating the Obese Body." *Communication, Culture and Critique*, no. 3, 2010, pp. 105–126.

Sidney, Stephen, Barbara Sternfeld, William L. Haskell, David R. Jacobs Jr., Margaret A. Chesney, and Stephen B. Hulley. "Television Viewing and Cardiovascular Risk Factors in Young Adults: The CARDIA Study." *Annals of Epidemiology* vol. 6, no. 2, 1996, pp. 154–159.

Smith, P. "Hooked On-line! Internet Costing People their Jobs, Homes and Families." *National Enquirer*, 1996, April 23, p. 11.

Standage, Tom. *The Victorian Internet: The Remarkable Story of the Telegraph and the Nineteenth-century's On-line Pioneers*, Walker Publishing, 2007.

Stewart, S. "Fly-eyed Baby!" *Weekly World News*. 2007. http://weeklyworldnews.com/mutants/225/fly-eyed-baby/. Accessed May 19, 2011.

Swerdlow, S. A. *13-Cubed: Case Studies in Mind-control and Programming*, Expansions Publishing Company, 2008.

Terrence, A. "Television is Poisoning American Minds Says Robert Montgomery." *National Enquirer*, 1969, June 18, p. 8.

"Too Much TV Watching Leads to Malnutrition." *National Enquirer*, 1969, December 7.

The Tube. BBC documentary. Dir. Trozzo. 2003, UK.

Tuckett, Kate. *Conspiracy Theories*, Penguin, 2005.

Wald, Gayle. "'I Want It That Way': Teenybopper Music and the Girling of Boy Bands." *Genders*, no. 35, 2002.

Walter, Richard D. "What Became of the Degenerate? A Brief History of a Concept." *Journal of the History of Medicine and Allied Sciences*, vol. 11, no. 4, 1956, pp. 422–429.

Wee, Valerie. 2011. "Visual Aesthetics and Ways of Seeing: Comparing Ringu and The Ring." *Cinema Journal*, vol. 50, no. 2, pp. 41–60. doi:10.1353/cj.2011.0003

Weheliye, Andrew G. "'Feenin': Posthuman Voices in Contemporary Black Popular Music." *Social Text*, vol. 20, no. 2, 2002, pp. 21–47.

Weheliye, Andrew G. *Phonographies: Grooves in Sonic Afro-modernity*, Duke University Press, 2005.

Weiss, A. S. *Phantasmic Radio*, Duke University Press, 1995.

"What's in the Box?" Dir. R. L. Bare. *The Twilight Zone*. 1964, USA.

X, Commander. *Mind Stalkers: UFOs, Implants and the Psychotropic Agenda of the New World Order*, Global Communications, n.d., ads suggest 1999.

· 7 ·

TECHNOPATHOLOGIES AS OUTBREAKS

Carriers and Demonized Collectivity

> Just think of those shocks you've got
> And those knocks you've got
> And those blues you've got
> From that news you've got
> And those pains you've got
> (If any brains you've got)
> From those little radios.
>
> —Cole Porter, "Anything Goes," 1934

> "Really? The Claim: Cellphones Can Cause an Allergic Reaction."
>
> —*New York Times*, 2010

A cough in the dark opens the medical thriller film *Contagion*. Sickness is the first thing signified. It is then shown to emanate from a woman, played by Gwyneth Paltrow, at an airport bar. She talks on her phone to a man with whom she had sex the previous night. The camera follows as she transfers her sickness from her hands to a bowl of peanuts to a credit card to a server's hands to a credit card reading machine. The title for the scene is "Day Two," and, as the movie unfolds, she and millions of others around the world will die from a deadly pandemic.

At end of the film, the threads of the disease's origin come together. A pig infected by bat droppings is sold to a restaurant in Hong Kong, where a chef prepares the carcass. Paltrow's character, a traveling businessperson dining at the restaurant, asks to personally thank the chef. As he leaves the kitchen to satisfy her request, he wipes his hands on the apron but does not wash or sterilize them. Shaking her hand, he spreads the disease to her. "Day One" proclaims the title, and the film ends.

Even though the male chef is the carrier, the film clearly positions a woman as Patient Zero, the source of the epidemic. The film does not tell us whether or not the chef gets sick; the narrative is solely focused on the chain of events rising from Paltrow's character. She is the conduit of the disorder, the medium through which pollution crosses from animal to human realms. She is also morally corrupt for, as we learn during the film, her sexual partner was an ongoing adulterous affair. The rendezvous that led to her spreading the disease to him is something she initiated by calling him and changing her flight times to accommodate a tryst.

In other words, the film's etiology of disease lands squarely on a morally degenerate female. Her heroic husband, played by Matt Damon, is, in contrast, immune to the disease. This perpetuates not only the stereotype of feminine pathologization, but also associations with degeneracy and a technological dimension: global mobility. She is not only a globe-traveling businessperson, she is also pointedly shown using mobile communication technology to arrange the encounter that triggers the pandemic. The other technological dimension is the credit card and reader: infrastructure of global capitalism during neoliberalism.

This chapter focuses on technopathologies in which unhealthy users withdraw from society. Here my argument complicates the earlier discussion of the individualization effect of technopathologies by counter-narratives, I argue, that are ultimately about demonized collectivity. Focusing primarily on cases related to mobile personal audio players, I apply Priscilla Wald's analytic of contagion to examine these as outbreak narratives. I will argue such sick users are pathologized for not merely self-isolation, but more for their potential reorganization into new social structures. Withdrawal is not merely the act of being antisocial. Disengaging from existing social structures is a first, necessary step in social reconfiguration, and therefore a polluting threat to the social status quo. I will present this first through an analysis of media coverage of Trisha Meili, aka the Central Park Jogger, victim in the infamous "wilding" gang-rape case, whose narrative shifts over time from outbreak victim to a carrier of threatening feminism and gender nonconformity, with her portable stereo worn when attacked becoming a significant component. I will support this analysis with other examples of similar technopathologies linked to other threatening social reconfigurations, such as race, HIV/AIDS, and juvenile delinquency. Following the discourse to its limit—hordes of audio monsters—underscores the cultural work of these discourses in demonizing collectivity. The contagious threat of sick, withdrawn users is not their social disengagement, but the potential outbreak of new social formations.

Outbreaks and Moral Contagion

Mary Douglas notes that, drawing on Durkheim, contagion is not inherently bad, but the dilution of the sacred. "The sacred must always be treated as contagious" in part because it is dangerous to cross "forbidden boundaries."[1] The sacred refers not merely to religion, but faith as well as other foundational aspects of social order. In her famous anthropological analysis of pollution taboos, contagion represents the spread of unclean social "dirt," and "dirt is essentially disorder."[2] From a pathological perspective, this is particularly relevant to disease. The order and boundaries of an individual body symbolize those of the larger social body to which it belongs. Contagion crosses boundaries and sows disorder; it represents the spread of threats to normative categories of same/different and self/other, thereby reinforcing disease as far from a neutral state.[3] Social cohesion is maintained by rituals of protecting against pollution and cleansing it in order to reinforce social structures, particularly the gender binary. Importantly, clean is not the opposite of dirty, or in the case of this book, sick is not the opposite of healthy, but often more an issue of degree: "a semblance of order is created" by exaggerating differences, not creating opposites.[4]

If contagion, then, represents a threat to social order, Priscilla Wald's analysis of recent contagion discourses is useful here. Examining typhoid, HIV/AIDs, virology, and other topics of science writing, news, and entertainment, Wald identifies an "outbreak narrative" which can apply to various dimensions of physical, social, moral, and emotional contagion. The conventions of this narrative involve

> a formulaic plot that begins with the identification of an emerging infection, includes discussion of the global networks throughout which it travels, and chronicles the epidemiological work that ends with its containment. As epidemiologists trace the

routes of the microbes, they catalog the spaces and interactions of global modernity. [This results in] a contradictory but compelling story of the perils of human interdependence and the triumph of human connection and cooperation, scientific authority and the evolutionary advantages of the microbe, ecological balance and impending disaster.[5]

Wald's analytic is useful here as electric communication technologies not only gird such global networks, but are means of "human connection and cooperation"—indeed, perhaps a most succinct definition of communication itself—but also its contradictory, dystopian incarnation in technopathologies. Moreover, as Tomes remind us, such disease narratives become cultural commodities, all the more appealing in their power to generate profitable news and entertainment.[6] There is even a significant body of historiography that posits disease as a driving force in historical change and social shaping, in a sort of pathological determinism.[7]

The film *Contagion* fits clearly within this narrative and, as Wald demonstrates, many elements of it are not so new. One of the most famous Patients Zero was Gaëtan Dugas, a gay French Canadian flight attendant, who was suggested or accused by a variety of medical, gay, and popular sources, including the Centers for Disease Control and Prevention, of bringing AIDS to North America from Africa. Dugas was portrayed as rabidly promiscuous and sociopathic, indifferent to his impact on others, even when he knew he was sick. In short, he was a feminine (gay, hysterical, sick, promiscuous, French Canadian, flight attendant), pathological (sociopathic, amoral, infected), and mobile (international flight attendant) degenerate who linked animal and human realms. Since discredited, numerous websites today still perpetuate the story that he caused the AIDS pandemic by having sex with a monkey in Africa, the disease a racist fantasy of homosexual miscegenation.

Identification: Technopathologies of Social Withdrawal

The pollution that threatens the social order in this chapter is withdrawal, the removal of individuals from proper social engagement and participation. In these technopathologies of withdrawal, instead of harnessing communication technologies to *connect* with others, in the grand tradition of American Progressive thought, the user *isolates* his or herself. These users withdraw into technologies to escape from the real world, evidencing pathological behaviors of antisociality, degraded perceptual capabilities, addiction, sociopathy, catatonia, dissociation, and hallucinations.

The outbreak narrative here is not a specific microbe, but the more general "emerging infection" of excessive withdrawal from social structures. This social or moral pollution is then traced through global networks of communication technologies, which are the focus of the narratives. Across these narratives, there is no unified content that affects the user: It is not specifically the seductive rhythms of rock and roll, the force of electric demons, or the hyperstimulation of video games. The sickness however, is the withdrawal into these various forms of content carried by electric communication technologies.

These occur across many forms of electric communication technologies. A long tradition of social science research has attempted to confirm degrees to which computer-mediated communication negatively impacts relationship forming and maintenance and causes Internet addiction and isolation—to mixed and inconclusive results.[8] Today, we continue to see cases in which withdrawal into the Internet has been blamed for people neglecting relationships, such that they lost jobs and families.[9]

Decades earlier, however, in "What's in the Box?"—an episode of *The Twilight Zone* television series—a doctor explained to a woman her husband's withdrawal into television:

> It's possible for anyone to have delusions, directly attributable to our over-mechanized culture ... Now, you say he's a TV addict. Perhaps he's been staring at this electronic ... television set for so long that its life has become his. ... And he's reached such a stage of confusion that he no longer knows whether he's watching the action or participating in it. Now, mind you, this is not an illness affecting only the juvenile or moronic. I, on occasion, have found myself asking for sutures or sponges during operations performed by television surgeons.

Fantastic narratives take withdrawal to an extreme of disappearing into immersive technological realities: the virtual reality systems of *Ghost in the Machine*, *Strange Days*, *Brainstorm*, and *Lawnmower Man*; entering TV shows and movies in *Shocker*, *Pleasantville*, *Willy Wonka and the Chocolate Factory* ("makes their brains as soft as cheese," chant the Oompa Loompas in reference to TV-addled miscreant Mike TeeVee), and *976-Evil II: The Astral Factor*; living within television studios in *Shock Treatment*, *The Truman Show*, and *Stay Tuned*; going inside video games in *Arcade*, *Game Over*, *Nightmares*, and *eXistenZ*; or computers, as in the music video, "Do You Computer?"[10] French photographer Antoine Geiger recently received press coverage for his manipulated photographs of public smartphone users' faces being stretched and drawn into the screens of their devices.[11]

For the remainder of this chapter, I wish to focus on audio communication technologies as a particularly rich vein of these discourses. A variety of scholars have argued that technologies of listening trigger multiple and recurrent cultural anxieties due to how they threaten public/private divisions, whether through broadcasting, mobility, or enabling women's participation in public realms.[12] An early version of this can be seen in a 1923 music company's advertising image titled "The Shut-in's Sunday Service" (Fig. 7.1).

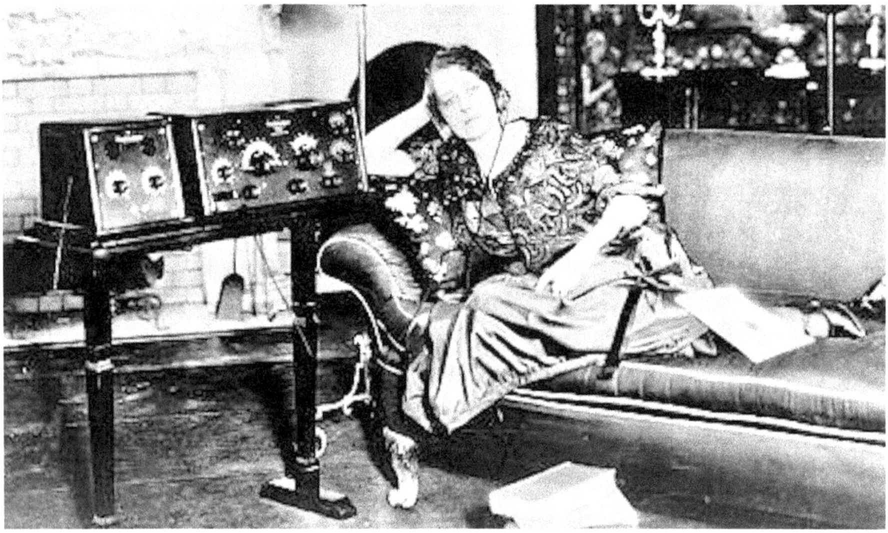

Figure 7.1. "The Shut-In's Sunday Service" illustrates the trope of the pathologized, feminized listener withdrawing into audio technologies.
Source: Clark Music Company, 1923. Library of Congress Prints and Photographs Division Washington, D.C. 20540 USA. Control #: 2004672638. Digital ID: cph 3c34575 http://hdl.loc.gov/loc.pnp/cph.3c34575. Original copyright Clark Music Co., J260769 U.S. Copyright Office. No known restrictions on reproduction.

Here, a literally sick woman is shown staring blankly while listening to a church service through radio headphones. She is wholly isolated, alone, with no visible windows, only a partial glimpse of a door, and even the fireplace bricked over. A mirror behind her reflects the floorboards of a largely empty room. This image can be viewed as a pathological extension of the emergent "private listening" cultural historian Jonathan Sterne describes as a new form of listening associated with audio technologies.[13] Bill Kirkpatrick has shown how this image, as part of larger discourses of the shut-in, played a significant role in harnessing norms of ability and disability in shaping federal broadcasting policy towards radio as a medium for passive audiences.[14] Indeed, the stereotype of the frail, feminized loner who is privately absorbed in phones, record players, or radios, is a potent one, which has appeared in decades of films produced in multiple countries, from *Nightmare* and *Little Voice* in the UK to *Les Yeux Sans Visage* in France and *Txt* in the Philippines. In Gian-Carlo Menotti's 1947 short comic opera, *The Telephone*, a woman is so enraptured by her telephone that her boyfriend has to leave the house and call her in order to gain her attention long enough to propose.[15]

Audio withdrawal can also make the private listener vulnerable to danger: A wired telephone conversation blocks perception of a killer's approach in *Don't Answer the Phone!*, and headphones on a home stereo and audio electronics hinder awareness of immediate threats in *He Knows You're Alone*, *Leviathan*, *Poltergeist*, and *The Wave*.

Global Networks: Mobility

Portability affords mobility, a necessary component for outbreaks. Mobility, in turn, affords independence, which here becomes pathologized as independence from existing social structures—and potential reorganization into new ones.

One of the first widely adopted, portable electric communication technologies was the transistor radio. First successfully marketed in 1956, it became one of the most popular ECTs of all time, and was one of the first widespread experiences of technological mobility. As such, it combined aspects of mass-media broadcasting with increased personalization and individualization. Transistor radios became common symbols of the global spread of modernity. Reporters traveling to developing nations and isolated places frequently used the seeming incongruity of a local native with a transistor radio to signify the meeting of old and new, savage and civilized, and link mobile, individualized communication with modernity and social progress. For example, in a profile of American "educationaries" being sent to teach in Africa, one says, "'I'm trained in technology—something that the Africans need.'" Yet, the article also describes how "one can walk into a shop in Kampala, Uganda and buy the latest electronic equipment, such as tape recorders and transistor radios—and come out to see lepers in the street, or have trouble getting a glass of boiled water."[16]

In such cases, we see the reverse parallel of Wald's outbreak narrative. Whereas, typically, the threat is a pollution to social order that comes from uncivilized realms—whether they be microbes from savage Africa, promiscuity from a reckless gay man, or abstinence from an uneducated Irish domestic servant—here modern, civilized realms bring order to those realms. The transistor operates as a signifier of order, cleaning the local society.

Indeed, in the U.S., many at first hailed transistor radios as signs of social progress, given how they solved the "rock and roll problem" by taking the contentious music out of family living rooms and putting it into the shirt-pockets of youth. The radios brought order to uncivilized and often delinquent youth. However, particularly with the development of earplugs and headphones, the progress of individual mobility attached to a preexisting pathology of the isolated or withdrawn user.

Mobility joins, reinforces, and expands these types of technopathologies. Raymond Williams theorized the concept of "mobile privatization," in which technologies of communication and transportation contributed to individuals who were mobile yet "deliberately self-enclosed."[17] Such self-enclosure becomes additionally pathologized with mobile audio technologies, particularly for the way in which they allow bad users to consume more streams of stimuli than they are able to safely process. A woman in one of my focus groups pathologized herself when describing a moment of audio withdrawal in which she lost awareness of the device itself: "One time I was talking on my cell phone, and I was going crazy. I was looking for my cell phone, and I was talking on it at the same time."

Furthermore, their situational unawareness makes them vulnerable to injury or death. The user pays too much attention to the wrong thing. A conversation in one of my focus groups suggested this when a male participant related how it was his car radio that had prevented him from hearing the warning sirens of a police officer. Another agreed, "You can't hear sirens. I mean, I'm sure if we did a survey, a lot of people are getting hit running around with headphones because they don't hear nothing, no sirens or anything."

In numerous films, a user is rendered unaware of potential dangers because he is listening to music on portable headphones. For example, a headphoned driver misses an urgent warning shouted at him in *La Balance*, a babysitter is nearly shot by a child psychopath in *Bloody Birthday*, and, in *Graduation Day*, a jogger sporting bright red headphones does not hear the killer come up from behind her. Her murder sequence ends with a slow pan from her headphones, strewn across the path and blaring music, to her dead body, emphatically linking the two. A female focus group participant noted the same thing in true-crime programs and forensic documentaries, and another participant described the pattern as, "Usually a lot of them have earphones on. They're jogging or whatever, and they can't hear anything, and they get killed." However, they did not feel this was only in media representations. One woman related,

> I think it's a dangerous situation when you do have earphones, especially if you're jogging late or depending where you're jogging, in a wooded area. ... It's just too dangerous to do that, especially for us women. Yeah, I think they do make themselves vulnerable for attack, you know, potential rape.

Withdrawn listeners on portable audio technologies are a concern in and of themselves, but their relation to outbreak narratives of social pollution becomes more explicit when they are positioned as threats to others. For one, they cannot hear when others are in danger. A woman in my focus groups described disregard for others as a result of listening on headphones: "You're completely oblivious to the world around you; when someone's in trouble you can't help them." Scholars have described withdrawal from nearby social connections a hallmark of mobile telephone use.[18] Again challenging the newness hypothesis, six years after the debut of the first digital audio player, there was nothing anxiously new about a Walkman cassette player blocking the sound a friend's nearby death in *A Soldier's Girl.*

Heightened vulnerability includes simple accidents as well, in a technological re-creation of the archaic medical condition of being "accident prone." Here, the exacerbation of inherent clumsiness reinforces the individualization effect of locating the technopathology within the user. A female focus group participant confessed,

> I'm a clumsy person. I used to drop my cell phone all the time. I would always just like get down and grab it until I heard a story of a guy who did that and then got in a car accident and died. So, now I don't, but I always get distracted by my cell phone like I always text and drive, and I drive a stick, so it's not good.

A male participant related that his daughter fell and hurt herself from "running around the house" with a handheld video game. Another participant, male, related:

> I got into a car accident once with one of my exes. She was talking on the cell phone, and I couldn't hear the person. She was like, What and what? And for some reason

people look down like as if the answer is on the ground. And I'm like, Stop, stop! And she slammed into [an object]. We were both fine, but I still have neck problems.

The withdrawn listener is pathologized as not merely vulnerable and dangerous, but, at times, overtly antisocial, when their self-absorbed focus on technology use rudely disregards others and breaks down established boundaries of public and private space. In the 1980s and 1990s, personal stereos—portable cassette players, then CD and minidisc players—were associated with "electronic narcissism,"[19] "ecstatic alienation,"[20] or "Walkman effect," in which the device "intrudes inside the skin ... The surface tension of the skin loses its balancing function through which it activates interpenetration of self and world."[21] Responding to a comment about technologies making users dependent and unthinking, a woman in my focus group said, "The isolation leads to self-absorption, and that makes people inconsiderate of others." More recently, private listening and public talking have been castigated in mobile phone use, when callers disregard others who are in earshot. This pattern has been noted not only by mobile phone scholars but also in articles and cartoons about "cell phone victims."[22]

Containment

While personalizing the pathology as individual, not systemic, the individualization effect also warrants monitoring and control of those pathologized persons for the safety of others. Following Wald's outbreak narrative, here we see the institutions of democracy and modernity, such as science, journalism, and legislation, marshaled to fight the outbreak and attempt to control the pollution. In 1982, a New Jersey township attempted to ban drivers, bicyclists, and pedestrians at intersections from wearing the headphones of personal cassette stereos, such as the Walkman.[23] Later, this concern was also applied to mobile phones, whose use while driving was restricted in the U.S. as well as Brazil, Switzerland, Israel, Australia, and Japan.[24] With the advent of portable digital music players, again legislators attempted to prevent oblivious headphone wearers from walking into traffic. "Warning: Be On The Look Out [sic] For Killer iPods!" trumpeted an online technology column describing a story, also covered in mainstream print and broadcast media, about New York State Senator Carl Krueger's assertion that "iPod oblivion" caused users to fatally walk into traffic.[25] Krueger had proposed a bill banning use of portable electronic devices while crossing city streets.[26] In 2011, users could download Etymotic, a free smartphone app that piped emergency sounds into the headphones of pedestrians.[27]

Finally, users in my focus groups supported the individualization effect in that they rejected the idea of regulating portable audio use, or redesigning the devices with volume limits. For example, responding to a comment about video technologies being "hypnotic," a male member of my focus groups added, "Same thing about the phones, cell phones. People just get so into them. They just forget where they're at, especially when they're driving, [and] cause accidents." Even though the technology was hypnotizing the user, it was still the mobile user who "caused" the accident. In this and other examples, my participants felt, almost unanimously, that the responsibility belonged to individual users. Some participants said they lowered their volume or ran with only one earbud in; others related that they didn't bother, merely accepted the risk. This latter dimension suggests the ultimate normalization of disease I will discuss later.

The Central Park Jogger: From Outbreak to Carrier

Mobile audio technology plays a significant role in media coverage of the case of Trisha Meili, aka the Central Park Jogger, who initially appeared as a white victim, but later was represented as having been over-ambitious, gender-non-conformist, and a bad technology user. Revealed to have been wearing a Walkman, her attack is reframed as being in part due to her being over-stimulated, her degraded capacity for processing multiple streams of perceptual information linked to, not merely attention, but desire. After her rape and subsequent 13 years of anonymity, in a publicized autobiography and interview with Katie Couric, she is represented as less ambitious and more appropriately gendered, exemplifying how the pathological pattern is connected to other normative patterns.

As with Paltrow's character in *Contagion*, Meili is an attractive, upscale, heterosexual white woman who threatens to become a carrier for perceived threatening violence and savagery to threaten social order. The "thirdworldification" of Wald's outbreak narrative is a geographic point of origin in *Contagion*; with Meili, it is initially the race of her attackers.

An outbreak begins with its identification, which is not the beginning of its spread, but when the spread becomes widely visible and thereby identified. Urban racial tensions were not new in 1989, but they did explode into public awareness to a new degree with the Case of the "Central Park Jogger." A 28-year-old investment banker, later known to be Trisha Meili, was jogging at night in Central Park, New York City. She was brutally attacked, raped, bound, and left for dead, but found in time for a hospital to slowly nurse her back to health. The case became a media sensation, traveling through communication networks to push aside stories such as the Exxon Valdes oil spill and closing arguments of the Iran/Contra trial. The anonymous Jogger was chosen as one of *Glamour*'s "Women of the Year" and *People*'s "Year's Most Interesting People." Forces of containment were rallied, and police arrested several suspects from subsidized housing and housing projects, all Black and Hispanic boys, aged 14–16, their races linking the identified threat to global networks of immigration, colonialism, and slavery. Media reports of their confessions seized on the term "wilding" as their slang for recreational gang rape. The obsession with this term in the popular imagination treated it like some kind of linguistic virus, epidemiological proof of the threatening disorder or moral contagion. Moreover, their youth, races, and perceived slang positioned them a harbinger of things to come: a portrait of emerging social reorganization or pollution of the status quo. Demonstrating the "Africanization" fears of outbreak narratives, the case crystallized contemporary racial tensions and fears about urban decay and youth lawlessness. Expressing his opinion on the case, Donald Trump took out four full-page ads in major New York newspapers urging "Channel your outrage: Demand the Death Penalty."[28]

Leading up to and during the trial, community leaders and family members voiced concerns that, Meili being white, the prosecution and media furor were racially driven. Five boys were convicted, despite claiming at trial that their confessions had been coerced. Upon arrival in Rikers Island prison, inmates beat two boys.

Outbreak Revised: A Carrier of Feminism

Thirteen years later, in June of 2002, a convicted murderer, Matias Reyes, claimed from jail that he had been the lone rapist of the Central Park Jogger. DNA tests placed him at the scene and found no links to the convicted boys. On December 19, a New York State Supreme Court Justice vacated the convictions of the five young men, which was allowed by state law if new evidence arises that probably would have resulted in a more favorable verdict—but it did not necessarily mean that they were innocent. Meili and prosecutors remained unsure and still suspected that the boys somehow had been involved.

They could have gone to trial again, but decided against it. Later research suggested that the media hype over the neologism "wilding" had been incorrect and misleading: the boys had actually been quoting a then-popular Ton Lōc song, "The Wild Thing," not "going wilding."[29] In 2003, Meili revealed her identity and went public with her autobiography.[30] Media coverage included a primetime, hour-long special with NBC News' Katie Couric. That same year, the five men filed suit against the city for malicious prosecution, racial discrimination, and emotional distress. They become the subject of a film by famous documentarian Ken Burns in 2012, and two years later the city settled the case for $41 million, without admitting any wrongdoing.

What, then was the outbreak? More precisely, how was sense to be made of the previous outbreak narrative? A single rapist didn't carry the same threat of impended social reorganization as the group of young men. As I will describe, during the revision of this outbreak, a technopathology becomes central and is used to pathologize Meili as a sick user who is a carrier of social threats of feminism and gender nonconformity.

Meili had been listening to music on headphones while jogging, interfering with her perceptual abilities in a technopathology of withdrawal, as described previously. However, the newspaper coverage I have been able to find of this case at the time does not mention the headphones. This is in contrast to contemporaneous press reports that overtly criticized Walkman-wearing. For example, one month before Meili's attack, an article ran in the *New York Times*, Meili's hometown paper, about an attack on Tiffany Sessions, a Walkman-wearing female jogger.[31] It read, in part:

> "Don't wear a Walkman," said Spencer Mann, a spokesman for the Alachua County Sheriff's Department … "Three days after Tiffany disappeared, I was out driving about 8 P.M. in the same area where we think she disappeared," Mann said. "I saw three different female joggers wearing a Walkman and running alone. And this after tremendous publicity about this young woman who was missing in that area. It just blew my mind that they could be so unaware [sic]."

A month after Meili, another *New York Times* article described "stealing … personal items like Walkman tape recorders and bicycles from vulnerable-looking people on the street" as part of the normal and ubiquitous street violence of East Harlem youth gangs.[32] No mention of Meili. A 1991 *Washington Post* mugging article is similar.[33] A 1994 *New York Times* article, "Subway Rape Victim tries to Prove Agency Was at Fault" actually suggests the contrary, blaming the Walkman-wearing victim in its lede: "Adjusting the earphones of her Walkman, the young administrative assistant entered the subway."[34] In 1999, the *New York Times* did mention a Walkman worn by a male victim in a Central Park killing.[35] Moreover, in 1989, the Walkman and similar devices had great cultural salience in general, evidenced by films depicting the phenomenon of the headphone-wearing vulnerable attack victim.

Yet, despite being one the most infamous crimes in New York history, Meili's Walkman is missing until the 2002–2003 news stories, when it becomes a crucial piece of evidence in the District Attorney's report and news accounts regarding Reyes and the convicted boys. Then, details of the type of Walkman and how it disappeared were among several consistencies supporting Reyes' account and inconsistencies in the boys' confessions. These, combined with the new DNA evidence, led to vacating the convictions of the boys originally accused of attacking Meili.

It is nearly impossible to know with certainty why initial reports did not mention her Walkman, when the same and other papers *did* mention Walkmans in other cases before and after. Perhaps reporters and editors were rightly sensitive to the horrifying violence of the case and taking extra care not to blame the victim by faulting her for wearing headphones. Chancer argues that this case pitted gender against race.[36] Taking a perspective of intersectionality, it can be understood that initially Meili was more important as a *white* woman. Perhaps what made Meili unique was that her technology use implicated her in terms of gender—technology misuse and incompetency are often associated with femininity, in contrast to masculine technical mastery and aptitude. This had to be avoided to present a positive image of Meili *against* her attackers. In other words, blaming the white, successful, female victim in this case would have risked weakening the case against the dark, poor, young men.

This is further supported by the partial discrediting of Meili 13 years later, when the technology use *does* appear. Once someone *else* emerges as a more likely culprit than the boys who had been convicted, Meili becomes less the noble, white victimized female and more the hysterical woman. The culpability is supported by the original absence of the Walkman detail in the media as well as from Meili's later autobiography. The Walkman does not make her look good; it makes her a bad user of technology. However, by 2002–2003, fears of the racially tinged New York crime crisis had lessened, due to Mayor Giuliani's crime war and gentrification of the city. Furthermore, a postfeminist cultural sensibility had become more entrenched, in which perhaps editors were no longer as concerned with being perceived as blaming a female victim. Indeed, the case is reframed as a story of wronged justice in Ken Burns' documentary for PBS, *The Central Park Five*. In her 2016 campaign, former Secretary of State Hillary Clinton brought up her opponent Donald Trump's refusal to acknowledge the innocence of the Central Park Five as evidence of his racism.[37]

Containing the Carrier

In the reframing of the case, Meili's story shifts from that of an outbreak's victim to one of the containment of a carrier. Technopathological discourses are deployed to demonize the social threat she represented: feminism and gender nonconformity. Containment here is not epidemiology and the triumph of science, but the triumph of "nature" over disorder. I amend Wald's model here to suggest that, while modernity can be marshaled to triumph over nature, nature likewise can triumph over unnatural threats to the social status quo. Technopathological logic works in multiple directions: just as the transistor radio could represent the containing forces of civilizing modernity, it could also be a conduit for threatening moral and racial contagion. Similarly, nature, in the sense of the biological world, can be the contagious outbreak that threatens society in the forms of a microbe. However, I argue that nature, in the sense of an imagined natural law or order, can also be marshaled to contain the threat of moral contagion arising from the corrupting effects of society. The first is a modern understanding, the second romantic. However, unlike Cohen or Kittler, I assert these are simultaneous, ongoing entanglements, not successive epochs. Both Paltrow's character in *Contagion* and Meili represent a threat to social order in the forms of female empowerment, ambition, and general nonconformity, but in their romantic outbreak narratives, a racist, sexist "natural order" reasserts itself to contain them through death and rape.

The absence of the Walkman detail in early coverage of Meili supported her signifying the status quo: upwardly mobile heterosexual whites who felt threatened in New York City. Moreover, when the Walkman detail emerged and Meili was positioned as a bad user, she represented that time a *threat* to the status quo: feminist challenge to gender roles and economic and social hierarchies. This, I argue, is the true threat of the sick user who withdraws. Technopathologies demonize users who withdraw, but particularly in a way that preempts the potential social reconfigurations that such withdrawals might make possible. In this way, they not only align with certain stereotypes and hierarchical social categories, they also blunt certain attempts to challenge them by demonizing collectivity.

This is most evident in one of Meili's biggest media appearances, the Katie Couric interview, which called attention to Meili's headphones at its onset.[38] After a brief introduction of the case, Meili's first appearance (not seen, but heard reading aloud) stated, "I am that woman, until now known only as the Central Park Jogger. And this is my story." To which Couric's voiceover replied:

> She liked to listen to music when she ran. Billy Joel especially would help her keep pace. An investment banker, she ran to release the stress of 12-hour days on Wall Street. She ran to maintain the slender, muscular physique of a former ballerina. And she ran five nights a week, because despite the danger, she says, there was something appealing about jogging after dark.

The music was not mentioned again in the interview, yet its prominent, pointed, early reference defined the event. It was the first detail mentioned, with even the specificity of naming what music she preferred. Couric chose to put her regular, newscasterly vocal emphasis on the word "music." Couric linked this immediately to a pathological description: She was stressed out, a workaholic investment banker in the 1980s stereotype of Wall Street greed and pathological over-ambition. She was obsessive about her body—the five-times-a-week running will later be described as "compulsive." The ballet history alluded to stereotypes of a high-strung feminine nature, obsessive with eating disorders, which, indeed, it was foreshadowing. The interview would later make Meili's history of anorexia explicit. Even in her autobiography, she described the disease since age 15 as having some "deep connection" to her "compulsive need to run."[39] In the interview, this then linked to further suggestions of pathology in her irrational jogging at night. She did not stay within appropriate boundaries of safety (as if it was her responsibility, not the problem of the city to keep the park safe), exhibiting reckless disregard for danger simply because of an unnamed, unexplained, "something appealing" about it. She went outside boundaries in her pursuit of desires. In very few words, the beginning of the interview associated technology misuse with an overambitious, reckless, pleasure-seeking, and sick woman, gluttonous and over-desiring. Her attention, as directed toward goals and objects of desire, was split over not only too many channels of information, but too many *wants*.

These connections were made not solely through words. The sensory mismanagement of stimuli was further emphasized by cinematic techniques that put the viewer in the subjective point of view of Meili. Much as the innovative way the film *Possessed* made the audience directly experience the mental illness of Joan Crawford's character, the opening of the Meili interview attempted to make the viewer experience the impaired faculties of the headphone-wearing jogger. This opening sequence's illustrative, establishing shots of lights, night sky, jogging feet, and pavement alternated with and dissolved in and out of perceptual clarity. Some were in focus, some not. The depth of field changed. There were temporal shifts from present (Meili's book) to past (jogging, path). The "Billy Joel" line was spoken over a dissolve from the path to extremely out of focus, abstract traffic lights. Literally we saw, at the discussion of listening to music, a clear, sharp jogging path turn into a warning system that we could not clearly apprehend. The overall effect was disorienting, suggesting unstable memory, consciousness, or perception. This was heightened by spare, ominous music and sounds of rapid breathing, an anxious panting that could be merely from jogging but also sounded like panic. It underscored Couric's reference to "jogging alone," and the ominous sounds connected music to the naïveté of jogging alone at night. When Meili said, "Nothing could harm me," the screen faded to black. Ominous string music reappeared as the interview faded up from black to a spooky shot of the moon behind bare tree branches, focus and depth of field shifting. Over this, Couric contradicted Meili: "Until the night. …"

The opening sequence of the interview linked listening to music on headphones as a disorienting, defining element of the events of that night. Music, Walkman, or headphones were not specifically mentioned again, yet I suggest they remain in memory, so that when the danger of jogging alone *was* suggested; the further indictment of "and wearing headphones" needed not be spoken. Note Meili's defensive response later in the interview:

> Couric: When this happened, some people said, "Running alone in Central Park at night? That girl must be crazy."
>
> Ms. Meili: That question bothers me a bit, because this—the implication is, "OK, it's my fault that—that I was raped and beaten and almost killed?" These random acts, unfortunately, happen all the time.

This view was supported by one of my focus groups who, when discussing headphone technopathologies in general, spontaneously remembered the Meili case. As participants discussed the details of the case they could recall, several blamed the user: One suggested that the problem was not the device but how aware the person wearing it was: "If you're aware of your surroundings, then you can hear them running up. If you're not and you have something in your ears, you're not." One man mis-recalled her as dating a U.S. senator, thinking perhaps instead of Chandra Levy, the murdered aide who had been having an affair with Congressman Gary Condit. Another man suggested that stereotypical feminine vulnerability was the problem, not the technology: "A woman is obviously a better target than a man typically because they're more vulnerable, but I don't think that wearing the headphones makes them more vulnerable."

Bearing in mind the unspoken headphones, note now how the following section of the interview moved to associate Meili's career ambition with the disease of anorexia and her seemingly irrational regard for her own safety. The infection of unbounded, desirous ambition spread.

Couric:	In 1986, she landed at Salomon Brothers, so cutthroat, people would say, no one would stab you in the back, they'd come through the door with a hatchet. But not Trisha.
Mr. Vermylen:	She stood out among her immediate peers and—and really across the whole investment banking department as being very friendly, always with a ready smile, always cheerful.
Couric:	Running helped her cope with the pressure and with the secret struggle. From the time she was 15, she suffered from anorexia.
Ms. Meili:	I was very conscious of what I ate, and how much I exercised, and that was part of the compulsiveness about running, that I always had to be running.
Couric:	Why was it so important to you?
Ms. Meili:	Because I was feeling pretty average at work. I wasn't any speed demon or, you know, or track star, but I felt a lot of peace and a—I guess a sense of accomplishment when I was out at night. And there were other people out there, too, but usually not too many.
Couric:	Were you ever afraid to run at night?
Ms. Meili:	It's hard to hear me say it, I guess, but no, I wasn't. I wasn't afraid, because I had done it for a long time and nothing had happened. And so I had some sense of security from that, and I also had this feeling deep inside that nothing's going to happen to me.
Couric:	So not even patches of the park creeped you out?
Ms. Meili:	I did have some ground rules. I would never go to the north end of the park, the northernmost end of the park, but, you know, I always did go up to the 102nd Street crossdrive, always at the first part of my run. I suppose I did realize that yes, it's more secluded, but I'll just get it over with more quickly and then I'll be fine.
Couric:	Wednesday, April 19th, 1989 might have been a forgettable day. …

Elsewhere, her jogging at night and ambition were clearly criticized:

Couric:	Phi Beta Kappa, just one on a roster of honors. You went to Wellesley, two master's degrees from Yale, including an MBA. … You really were very impressive academically.
Ms. Meili:	I was a very disciplined student. You know, part of it was that I always wanted to be the best.
Couric:	At first she wanted to change the world, and considered joining the Foreign Service. But in bullish 1980s America, Wall Street was the place to go.
Ms. Meili:	I thought, well, let me just give this a try. Let me see if I can compete, you know, with, you know, with the best and the brightest.

Her tomboy childhood underscored gender nonconformity. Anorexia emphasized her pathologization. The targets of her ambitious competition ranged from childhood brothers to the boys' club of 1980s Wall Street—a path chosen over the more traditionally feminine diplomatic corps of Foreign Service.

The absence of the Walkman detail from initial media accounts and Meili's autobiography suggested self-consciousness, an awareness of the dynamics of blaming the victim. Yet, in later accounts from 2002 to 2003, we see not only the Walkman but there was ample blaming of the victim as over-ambitious, unbounded, and pleasure-seeking. And, in turn, this pathologization of the individual neglected larger, social factors that could have been involved. Systemic discussions were missing: Park lighting? Police patrols? The poverty of the youths involved? Why did they not have any place better to go that night? (Even with the outcries of racism during the trial, discussions tended to look at racism in terms of individual prejudices, not systemic inequalities and inequities.) Similarly, in the Tiffany Sessions case cited earlier, it did not "blow the mind" of local law enforcement that listening devices were designed in such a way as to create sense-impairing volume in the first place. Or that we, as a society, have men lurking in bushes waiting to pounce on coeds. The patterns of bad user, bad female, and bad victim all reverberated off on another, and worked to obscure larger, social patterns. These warnings for female joggers to avoid headphones continue to the present day.[40]

The technological pathologization was thereby heavily gendered, in a pattern we saw previously in the camera pan to the murdered, female, headphone-wearing jogger's signifier of sexuality—her curvy ass—in *Graduation Day*. Both were sick users in that they had their perceptual defense systems dangerously divided, making them vulnerable and susceptible to threats if not managed properly. Yet, particularly in Meili's case, the choice to consume multiple streams of sensory information was clearly articulated as parallel to gender issues: she is also too ambitious and too vain. To stay thin and release stress from trying to compete with the boys, she ran compulsively at night, wearing headphones. It was not her fault that she was attacked, but the implication clearly remained that if she hadn't been doing those things, she wouldn't have been attacked. The chain analysis was clear. Moreover, her punishment was the most heavily gendered imaginable: rape.[41]

Couric's concluding words to the interview made these factors clear: "Trisha still loves to run, but once a week now, no longer every day. Gone is the hard-charging professional trying to prove herself on Wall Street. But for all that was lost, she says, she's found something far greater: The true measure of her success, the capacity to be generous and to love." Ambition was abandoned and safely relocated from the competitive boys' club of finance to a more feminine caregiving career as an inspirational speaker. Meili emerged as more appropriately gendered by the interview's conclusion, her newfound capacity for generosity and love suggesting that previously she had been more masculine, cold, and selfish.

More recently, a post for *The Modern Man* blog drew attention online regarding this subject. Titled, "How to Talk to a Woman Who is Wearing Headphones," it described ways for heterosexual men to get past the technological barrier and get the attention of attractive women wearing headphones.[42] Its sexism made explicit the gendering of femininity as receptive to male overtures, and headphones presenting an obstacle to such ideal femininity and, thereby, performance of ideal masculinity in hooking up with these women.

If the original threat to social order in Meili's outbreak narrative had been represented by violent young men of color, it becomes later represented by Meili's own ambition and gender non-conformity. Rather than modern legal and medical institutions, here more neoliberal institutions of self-promotion, celebrity, and entrepreneurialism are marshaled to contain the threat. The pollution, the disease, becomes more clearly welded to the individual rather than an outside threat.

This is the individualizing effect at work: It is not a problem of the technology, it is the problem and responsibility of the user, and, failing this, various social institutions to assert control and discipline. In Meili's case, the social beauty standards and ideals of femininity that required an ambitious female banker to run compulsively to stay thin and attractive were left unexamined. In cases of mobile headphones in general, it is rarely mentioned that a home environment so filled with labor—professional, family—and entertainment—computers, phones, game system—leaves the only place one can enjoy again the pleasure of concentrating on uninterrupted music with headphones is when working out. The *New York Times* belittled users in an article about the attempt to ban headphone-wearing at intersections, arguing that, "Only a few pedestrians would be protected from their own obliviousness by forcing headphone wearers to take them off when they cross the street."[43] Yet, what of a workload and information-saturated mediascape that require multitasking, such as listening to podcasts, voicemail, spoken email, or audiobooks while walking to stay informed and culturally literate within one's professional or social spheres? The blatant pathologizing of Meili showed how technopathological discourse is normalizing discourse. These are the wrong kinds of users, but, by focusing on sick *users*, we lose sight of what are sick-ening aspects of *society*.

Consumption, entertainment, and stimuli-seeking all have a distracting effect. Jogging down the path, distracted by happy music, we do not hear the rapist approaching. But note the process by which the user is again to blame. How dare she wear a Walkman while jogging alone—it's almost like she was asking for it! And consider the double standard: we are constantly bombarded with incitements to do these exact things. Images of women out exercising alone with headphones on have been a hallmark of advertising, from recent iPod ads to the very first magazine ad for the Walkman, which showed a woman skating outdoors.

The Threat of Withdrawal: Social Reorganization

Death is not the only form of disorder. Particularly in the sense of moral contagion, disorder is any threat to the status quo, a potential social reconfiguration. Meili's threat was feminism. Focusing on the same technopathology of withdrawal through mobile audio, I will now show historic precedents in civil rights, HIV/AIDS, and urban youth, then extreme examples of techno-horror that explicitly demonize the potential collectivity and reorganization withdrawal affords.

To return briefly to the film *Contagion*: In it, Jude Law plays an independent blogger trying to expose truth behind the outbreak. Such technologically enabled muckraking could be viewed within Wald's outbreak narrative as a contemporary aspect of the institutional forces of social order mobilized against the threatening pollution, as well as the outbreak itself traveling through global communication networks—outbreak as meme. The narrative device of the investigator is traditionally heroic, from exposing government corruption in *All the President's Men* to corporate liability for health problems in *Erin Brokovitch* (also directed by *Contagion*'s Soderbergh). However, the character's depiction is villainous. While exposing some official wrongdoing, overall he is deceptive, spreading panic and false information about a cure from which he attempts to profit. He is arrested but freed when millions of his readers post bail.

This illustrates what I will now describe as the larger threat behind technopathologies of withdrawal. It is not merely destabilization of the existing social order, but the potential for then creating new social formations. As such, pathologized withdrawal actually demonizes collectivity. In the case of *Contagion*, Law's character represents the new social formations of citizen journalism not as an unhinged individual but a larger, threatening social collective. I will now discuss how such demonized collectivity is at play in technopathological discourse.

Numerous examples of technopathological discourse illustrate this, particularly with the specific technology of mobile audio.[44] The alienating, antisocial, atomizing, and distracting psychological states attributed to the Walkman personal stereo entangle with larger social anxieties around youth and social class.[45] Significantly, the Walkman was designed for and target marketed to a collective identity from the beginning: youth.[46] It is not surprising, then, that a technology associated with a disempowered social group would be so frequently castigated. Before the Walkman, portable radios were associated with degenerate youth in films such as *A Place in the Sun*, *Blackboard Jungle*, and *Crazed Fruit*. A *Wall Street Journal* article on transistor radios from 1956 called them "a boon for teen swains … Transistor radios … [are] not so apt to run down the car battery when played long hours with the auto parked and the engine turned off."[47] Shortly thereafter, *Life* magazine pathologized teenagers with transistor radios as "hooked on sound," "transistor addict[s]," "bleatnik[s]," and "red-eyed little tykes."[48] As technologies changed, pathologies remained the same: A *New York Times* article on teen drug dealing and use warned in its headline, "Ask Not for Whom the Beeper Tolls."[49]

The progression from withdrawal to threats of social reconfigurations can be suggested when isolated users are associated with those very groups that would represent a social threat if they collectively reorganized, such as racial and sexual minorities, economic underclasses, immigrants, women, and youth. For example, an ethnographic study of "Negro Teen-Age Culture" in the *Annals of the American Academy of Political and Social Science* noted "a universal interest in music, which seems to work [upon them] like a narcotic" creating an "intense" "craving" for stereos, records, and transistor radios, often driving them to steal to support the habit.[50] A 1965 *New York Times* feature used technology to mark new delinquent youth occupying a neighborhood: "They pack the numerous pizza stores lining the Concourse, all the time listening to rock 'n' roll and shutting their eyes and swaying as though hypnotized whenever their favorite song blares … from transistor radios." In describing the neighborhood's change, a social worker is explicit about the race of these youth, explaining that "'Negroes are moving onto the side streets and a lot of people who aren't admitting it are just plain frightened'."[51] A 1976 editorial even described the defiant blasting of transistor radios on public busses by young "rebellious" blacks as part of "insidious evil" that perpetuated their own oppression.[52] During the debate around an early ban on Walkmans and headphone radios in intersections, a columnist for the *Washington Post* expressed the personal pathologization he felt wearing his Walkman, describing hostile glares from passers-by and being told by his boss that he looked like an idiot. "They hate Walkmen. Lots of people do." Yet he concluded his defense through contradistinction with bad users of a different age and suggestively, race, positing as the bad example youth carrying around "elephantine radios" blaring music everyone else could hear, what were often called at the time "ghetto blasters."[53] A *New York Times* article described "stealing … personal items like Walkman tape recorders and bicycles from vulnerable-looking people on the street" as part of the normal and ubiquitous street violence of East Harlem youth gangs.[54]

The *Washington Post* associates pathological use of portable stereos more specifically with working-class African-Americans.[55] During the debate around an early ban on Walkmans and headphone radios in intersections, a writer for the *Washington Post* wrote, "For a brief moment in history, it looked like the onslaught of barbarians bearing boogie boxes (those gigantic radios) might be silenced by the arrival of the biggest street corner status symbol since alligator shoes. Enter the Sony Walkman and its myriad look-alikes, with their wispy headphones hooked up to lightweight tape decks."[56] The class and racial allusions to working-class African-Americans are clear in "street corner" and "alligator shoes," associating one form of pathologization (deleterious technology misuse) with other, more familiar ones (class and racial stereotypes), as discussed previously.

Pathologization of sexual minorities has been technologically reinforced as well. A 1988 *New York Times* exposé on AIDS among "reckless and irresponsible" adolescent homeless prostitutes concluded with the tragic image of a boy showing off the new haircut, leather jacket, and Walkman a man had given him, while a social worker shakes her head, sadly noting his AIDS-suggestive weight loss and doubting the relationship would last a week.[57] Seven years later, the *New York Times* described how subway bathrooms had turned "creepy" after the removal of attendants:

> A man with a goatee and a Walkman who was in and around the bathroom for four hours on Tuesday told a reporter gravely: "You better stop asking so many questions, you're going to get killed." Since the attendants were dismissed, the police have made four arrests for public lewdness.

Note the swift association of degrees of degeneracy: public lewdness in the same breath as homicide.[58] More importantly, the Walkman was deemed a newsworthy detail—more relevant than others that could have been mentioned, such as clothes, hygiene, or expression—perhaps because of the common cultural association of technology with degenerates.[59]

I cite these examples at length to demonstrate again how, even within a narrow and specific type of technology—personal, portable audio—technopathologies align with existing social hierarchies: tech degenerates are sex degenerates are juvenile degenerates are black degenerates.

The Extreme Case: Monsters

The withdrawal and reorganization of social groups has been frequently presented in fictional narratives of normal individuals transformed into monstrous, violent, and often cannibalistic mobs, ranging from werewolf packs and witches' covens to vampire communities and zombie hordes. The last two have been particularly evident in the last two decades of popular culture. I turn now to specifically zombie- and zombie-like transformations, as they appear more often in technopathological narratives.

It is important to remember that a strict separation between fact and fiction is disingenuous—they are instead entangled. For example, on a post on a conspiracy message board about how wifi, mobile phones, and other sources of electromagnetic radiation tear apart DNA to cause genetic mutations, a commenter replied that such forces can create "mind control slaves," "zombify" and "stimulate Prozac saturated brain into Murder-berserK mode (e.g., Columbine, Virginia Tech). (How about face-eating zombies!)"[60]

Zombie-like metaphors are used to describe the catatonic self-absorption and disconnection from the real world exhibited by users of portable audio devices: "Walkpeople," "iPod people," "Text Walkers," "glassy-eyed," "plague-carriers" with "autism."[61] However, the most explicit examples appear in horror films and fiction. *Dr. Frankenstein on Campus* uses a radio-like "brain-wave box" to control people and animals as his zombie-like slaves. In *Creature with the Atom Brain*, an ex-Nazi mad scientist and assistant used radio-controlled atomic-powered zombies in his quest to help an exiled American gangster return to power. *Dr. Orloff's Monster* featured radio frequencies controlling a reanimated corpse to do a mad scientist's bidding. The Spanish cult zombie film *Let Sleeping Corpses Lie* presented the angry dead reanimated by the sound waves and "ultrasonic radiation" of a pest control system.

Godfather of zombie cinema George Romero was one of two persons to whom Stephen King dedicated his 2006 bestseller, *Cell*—note the biological, political, and technological entanglements in just that one-word title. In this novel, mobile phones spread an electronic plague that turns people into vicious, homicidal zombies, who then collectively work to transform others. Using cell phones, portable stereos, and Walkmans, the zombies recharge themselves and intentionally infect others, although these devices are later used to inoculate people as well. The struggle over social structure took place through communication technologies: "It was the cell phones … It was the cell phones, all right."[62] Inspired by *Cell*, as well as *The Signal*, the 2016 short film *Dawn of the Deaf* makes the social reorganization even more explicit when a sonic pulse turns the hearing population into zombies, elevating a previously marginalized group—the Deaf—to lead the fight for survival.

Through the demonization of individuals who socially withdraw, gaining the potential to create new social formations, technopathologies pre-empt certain challenges to the status quo. The threatening aspect of separation is its potential for consolidation of difference. The individualization effect personalizes these sicknesses; the demonization of collectivity further discourages conceiving of these sick individuals coming together to organize collectively. Demonized collectives could be class-action suits of injured users, consumer rights groups, or organized demands for user-centered design. Consider the consumer product safety movements, particularly in regards to automobiles. Changes were made to the design of the technologies (safety belts, air bags) and their use was regulated (speed limits, safety belt laws). Yet, as seen in my focus groups, options such as product re-design and usage regulation, whether brought about by grassroots political organizing, class-action lawsuits, were rejected outright as impractical, corrupt or amusingly oddball. At best they were considered only as minimal or limited alternatives to consumer education and individual changes in technology use.

When discussing preventative measuring for technology-related health risks, one participant suggested educational messages: "Yeah, the [communication technology] manufacturer, just like the cigarette company, is in charge in delivering their [health warning] message, not that it always works. I mean, it's up to the consumer at that point." Merely posting warnings so that it's "up to the consumer" was a common perspective, one that not only made health issues a personal responsibility, but completely erased government agencies and social activists—the actual parties responsible for forcing cigarette manufacturers to post health warnings. Moreover, posting warnings was seen as less protection for the consumer than the manufacturer. As another participant said, "We're warning you ahead of time. This way, you can't sue a telephone company or the product ... and blame them for it."

Notes

1. Douglas, *Purity*, 27.
2. Ibid. 2.
3. Ibid. 142; Shildrick, "Vulnerable," 155.
4. Ibid. 5.
5. Wald, *Contagious*, 2.
6. Tomes, "Epidemic."
7. Schell, "Sexist Gene."
8. Baym, "Interpersonal Life Online," 45–48.
9. Smith, "Hooked On-line!"
10. I exclude *Tron* and narratives in which electronic worlds are parallel or adjunct to reality, and their crossovers not pathologized.
11. McGinley, "Smartphone Suck".
12. Berland, "Musicking"; Chambers, *Migrancy*; Douglas, *Listening In*; du Gay, Hall, Janes, Mackay, and Negus, *Walkman*; Hilmes, *Radio Voices*; Keates and Scott, *Going Public*; Kruse, "Early Audio"; Sterne, *Audible Past*.
13. Sterne, *Audible Past*.
14. Kirkpatrick, "Shut-In."
15. Eaton, *Opera*.
16. Samuels, "Meet Africa's," 36.
17. Williams, *Television*.
18. Ling, *Mobile Connection*; Robbins and Turner, "United States."
19. Chen, "Electronic Narcissism."
20. Moebius, "Colouring."

21. Hosokawa, "Walkman Effect," 176.
22. Katz and Aarkhus, "Introduction"; Song, "Cell Phone."
23. "Consumer Saturday"; "Walkman v. Talkman."
24. Hafner, "Road Daze"; Kelley, "Suffolk Votes"; Kolata, "Researchers Warn."
25. Ryan, "Warning."
26. "N. Y. Lawmaker Hopes"; Rapoport, "New Yorkers"; Young, "Bill Banning."
27. Concotta, "Ear Gear."
28. Bailey.
29. Chancer, "Before and After"; Welch, Price, and Yankey, "Youth Violence."
30. Meili, *Central Park*.
31. Stockton, "Taking Precautions," C12.
32. Pitt, "East Harlem," B1.
33. Rubin, "Mugging."
34. Hoffman, "Subway Rape," B1.
35. Roane, "Body in Park," "Police Say."
36. Chancer, "Before and After."
37. Bailey.
38. "NBC Katie Couric Exclusive: The Central Park Jogger."
39. Meili, *Central Park*, 32.
40. "Video: How Women," Kumar, "On the Run Column"; Paiella, "Three Young Runners".
41. The pathologizing seen in Meili's case has continued elsewhere. A 2008 email broadcast to the entire student body of the University of Southern California did not specify the gender of the victims in two attack cases, merely described an "assault" on one student and another being "deceived into letting three strangers accompany that person home." However, a few pages later, the strong condemnation of "sexual crimes" suggested at least one was female and included a precaution "never wear earphones while walking alone" (Jackson, "Security Alert").
42. Bacon, "How to Talk".
43. "Walkman v. Talkman."
44. The use of visual communication technologies as part of a demonized collectives appears in the 1981 film *Dead and Buried*, in which murderous zombie hordes in small town obsessively take photographs and shoot home-movies as visitors to their small town are killed. The local mortician has developed a secret technique for reanimating the dead, and all of the townspeople are reanimated corpses under his control as he projects multimedia displays of multiple screens, films, and slide projections.
45. du Gay, Hall, Janes, Mackay, and Negus, *Walkman*, 89.
46. Ibid. 100.
47. Apple, "Transistor's Rise," 1.
48. Quoted in Douglas, *Listening In*, 226–227.
49. Roberts, "Ask Not."
50. Himes, "Negro Teen-age."
51. Weinraub, "Once-grand," 40.
52. Maleson, "Rebellious Blacks."
53. R. Cohen, "Walkman," AI.

54. Pitt, "East Harlem," B1.
55. Allen, "Battle."
56. Ibid.
57. Daley, "New York," B4.
58. M. F. Cohen, "Unattended."
59. Ibid.
60. BJ, "WiFi," Comment 7.
61. In the 2010s graphic-novel-turned-hit-television-series, *The Walking Dead*, zombies are called "walkers." See also du Gay, Hall, Janes, Mackay, and Negus, *Walkman*, 115, 118, 140; "Mayhem."
62. King, *Cell*, 45. Although maintaining the basic premise, the 2016 cinematic adaption lacked many of the technological aspects as well as the social reorganization of the zombies.

References

Allen, H. "Battle of the Ban: Can Woodbridge, NJ Uphold Headphone Law?". *Washington Post*, 1982, July 14, A1. Proquest database. Accessed June 14, 2007.

Apple, J. "Transistor's Rise." *Wall Street Journal*, November 29, 1956, 1. Proquest database (accessed June 14, 2007).

Bacon, Dan. "How to Talk to a Woman Who Is Wearing Headphones." *The Modern Man* 2017, June 12. http://www.themodernman.com/dating/how-to-talk-to-a-woman-who-is-wearing-headphones.html. Accessed June 15, 2017.

Bailey, Isaac. "How a Quarter-Century-Old Crime Became a Big Election Story in 2016." *Politico*, 2016, November 6 https://www.politico.com/magazine/story/2016/11/donald-trump-2016-central-park-5-214422. Accessed November 21, 2017.

Baym, Nancy. "Interpersonal Life Online." In *Handbook of New Media: Student Edition*, edited by Leah Lievrouw and Sonia Livingstone. Sage, 2006.

Berland, Jody. "The Musicking Machine." In *Residual Media*, edited by Charles R. Acland, University of Minnesota Press, 2007.

BJ. "WiFi: Invisible Threat That Pulls Apart DNA, Causing Genetic Disorders & Cancer." *Lunatic Outpost*. Comment #7, 2012, September 20. http://lunaticoutpost.com/showthread.php?tid=186172. Accessed September 22, 2014.

Chambers, Iain. *Migrancy, Culture, Identity*, Routledge, 1994.

Chancer, L. "Before and After the Central Park Jogger: When Legal Cases become Social Causes." *Contexts*, Summer, 2005, pp. 38–42.

Chen, S. S. "Electronic Narcissism: College Students' Experiences of Walkman Listening." *Qualitative Sociology*, vol. 21, no. 3, 1998.

Cohen, M. F. "Unattended, a Restroom Grows 'Creepy.'" *New York Times*, 1995, April 16, A1. Proquest database. Accessed June 14, 2007.

Cohen, R. "The Walkman." *Washington Post*, 1982. Proquest database (accessed July 23, 2007).

Concotta, Katie. "Ear Gear: An Escape from Reality or Death Trap?" *The Sydney Morning Herald*, 2011, November 10.

"Consumer Saturday: Headsets and Ear Damage". *New York Times*, 1982, July 17, 12. Proquest database. Accessed June 14, 2007.

Daley, S. "New York City Street Youth: Living in the Shadow of AIDS." *New York Times*, 1988, November 14, A1, B4. Proquest database. Accessed June 15, 2007.

Dead and Buried. Dir. Gary Sherman. USA, 1981.

Douglas, Mary. *Purity and Danger: An Analysis of Concepts of Pollution and Taboo*, Routledge, 2002.

Douglas, Susan J. *Listening In: Radio and the American Imagination*, Random House, 1999.

du Gay, P., Stuart Hall, L. Janes, H. Mackay, and Keith Negus. *Doing Cultural Studies: The Story of the Sony Walkman*, Sage, 1997.

Eaton, Q. *Opera Production I: A Handbook*, Da Capo, 1974.

Hafner, K. "Road Daze: A Hand on the Wheel and an Ear to the Phone." *New York Times*, 1999, July 29. Proquest database. Accessed July 27, 2007.

Hilmes, Michelle. *Radio Voices: American Broadcasting, 1922–1952*, University of Minnesota Press, 1997.

Himes, J. S. "Negro Teen-age Culture." *Annals of the American Academy of Political and Social Science*, no. 338, 1961, pp. 91–101.

Hoffman, J. "Subway Rape Victim Tries to Prove Agency was at Fault." *New York Times*, 1994, August 31. Proquest database. Accessed June 16, 2007.

Hosokawa, S. "The Walkman Effect." *Popular Music*, no. 4, 1984, pp. 165–180.

Jackson, M. "Security Alert from Vice President Michael Jackson." University of Southern California email to student body. Student Affairs, 2008.

Katz, James E., and Mark A. Aakhus. "Introduction: Framing the Issues." In *Perpetual Contact: Mobile Communication, Private Talk, Public Performance*, edited by James E. Katz and Mark A. Aakhus, Cambridge University Press, 2002.

Keates, D., and J. W. Scott, eds. *Going Public: Feminism and the Shifting Boundaries of the Private Sphere*, University of Illinois Press, 2003.

Kelley, T. "Suffolk Votes to Ban Cell Phones on the Road." *New York Times*, 2000, October 4, B1, B10. Proquest database. Accessed July 24, 2007.

King, Stephen. *Cell*. Scribners, 2006.

Kirkpatrick, Bill. "'A Blessed Boon': Radio, Disability, Governmentality, and the Discourse of the 'Shut-In,' 1920–1930." *Critical Studies in Media Communication*, vol. 29, no. 3, 2012, pp. 165–184.

Kolata, G. "Researchers Warn Driving and Phone Chats Don't Mix." *New York Times*, 1997, February 13. Proquest database. Accessed July 23, 2007.

Kruse, Holly. "Early Audio Technology and Domestic Space." *Stanford Humanities Review*, vol. 3, no. 2, 1993, pp. 1–16.

Kumar, Jay. "On The Run Column: Take Precautions When You Run." *The Daily News of Newburyport*, 2017, June 12. http://www.newburyportnews.com/sports/on-the-run-column-take-precautions-when-you-run/article_1d88edf8-01ba-519d-bbcb-0b41560fa98a.html. Accessed July 15, 2017.

Ling, Rich. *The Mobile Connection: The Cell Phone's Impact on Society*, Elsevier, 2004.

Maleson, A. "For Rebellious Blacks, It's the Bus to Nowhere." *Los Angeles Times*, 1976, Jan. 15, C5. Proquest database. Accessed June 5, 2007.

McGinley, Tara. "Smartphone Suck: Photos Depict Evil Screens that are Stealing Your Soul!" *Dangerous Minds*. 2015, November 10. http://dangerousminds.net/comments/smartphone_suck_photos_depict_evil_screens_that_are_stealing_your_soul. Accessed July 15, 2017.

Meili, Trisha. *I am the Central Park Jogger: A Story of Hope and Possibility*. Scribner, 2003.

Moebius, H. "Colouring the Grey Everyday: The Psychology of the Walkman." *Free Associations*, vol. 4, no. 32, pt. 4, 1994, pp. 570–576.

NBC Katie Couric Exclusive: The Central Park Jogger. 2003. [Television news special]. NBC News. USA.

"N. Y. Lawmaker Hopes to Ban iPod Use in Crosswalks." *ZDNet*. 2007. http://news.zdnet.com/2102-9584_22-6157109.html. Accessed February 7, 2007.

Paiella, Gabriella. "Three Young Runners' Recent Murders have Hit Women Runners Especially Hard, But Safety Precautions Have Always Been a Burden." *The Cut*. 2016, August. https://www.thecut.com/2016/08/women-runners-safety-vanessa-marcotte-karina-vetrano-alexandra-brueger.html. Accessed August 25, 2016.

Pitt, D. E. "For East Harlem Teen-agers, Posses Mean Violence." *New York Times*, 1989, May 9. Proquest database. Accessed July 23, 2007.

Rapoport, S. "Many New Yorkers Laugh Off Proposed 'iPod' Law." 2007. http://wcbstv.com/topstories/local_story_038164813.html. Accessed February 7, 2007.

Roane, K. R. "Police Say Teen-ager Died for His Sneakers and Beeper." *New York Times*, 1997, September 27, B3. Proquest database. Accessed July 23, 2007.

Roane, K. R. "Body in Park Identified; Killing is Suspected." *New York Times*, 1999, May 12. Proquest database. Accessed July 23, 2007.

Robbins, Kathleen, and Martha Turner. "United States: Popular, Pragmatic, and Problematic." In *Perpetual Contact: Mobile Communication, Private Talk, Public Performance*, edited by James E. Katz and Mark A. Aakhus, Cambridge University Press, 2002.

Roberts, S. "Ask Not for Whom the Beeper Tolls." *New York Times*, 1988, June 2. B1. Proquest database. Accessed July 23, 2007.

Rubin, A. J. "Mugging and Redemption." *Washington Post*, 1991, March 10. Proquest database. Accessed July 23, 2007.

Ryan, J. "Warning: Be on the Look Out for Killer iPods!" *The Post Chronicle*. 2007. http://www.postchronicle.com/news/original/article_21262969.shtml. Accessed February 7, 2007.

Samuels, Gertrude. "To Meet Africa's Great Need—Education." *New York Times (1857-Current file)*; August 20, 1961; ProQuest Historical Newspapers The New York Times (1851–2003), p. SM33. Accessed June 13, 2007.

Schell, Heather. 2002. "The Sexist Gene: Science Fiction and the Germ Theory of History." *American Literary History* vol. 14, no. 4, pp. 805–27.

Shildrick, Margrit. 2001. "Vulnerable Bodies and Ontological Contamination." In *Contagion: Epidemics, History and Culture from Smallpox to Anthrax*, edited by Alison Bashford and Claire Hooker, Pluto Press, 2002, pp. 140–156.

Smith, P. "Hooked On-line! Internet Costing People their Jobs, Homes and Families." *National Enquirer*, 1996, April 23, 11.

Song, L. "Cell Phone Victim—Aren't You Aware You Are One of Them." *Articlesbase*. 2009. http://www.articlesbase.com/online-promotion-articles/cell-phone-victim-arent-you-aware-you-are-one-of-them-1072195.html. Accessed May 19, 2011.

Sterne, Jonathan. *The Audible Past: Cultural Origins of Sound Reproduction*, Duke University Press, 2003.

Stockton, W. "Taking Precautions for Outdoor Exercise." *New York Times*, 1989, March 13, C12. Proquest database. Accessed June 27, 2007.

"Text Walker Mayhem." 2012. Allstate insurance commercial. http://www.youtube.com/watch?v=JnVSUYJYrD0&feature=youtube_gdata_player. Accessed June 15, 2015.

Tomes, N. "Epidemic Entertainments: Disease and Popular Culture in Early-Twentieth-Century America." *American Literary History*, vol. 14, no. 4, 2002, pp. 625–652.

Wald, Priscilla. *Contagious: Cultures, Carriers, and the Outbreak Narrative*, Duke University Press, 2007.

"Walkman v. Talkman." *New York Times*, 1982, July 13. Proquest database. Accessed May 13, 2005.

Weinraub, B. "Once-grand Concourse." *New York Times*, 1965, February 2, 35. Proquest database. Accessed June 14, 2007.

Welch, M., E. Price, and N. Yankey. "Youth Violence and Race in the Media: The Emergence of 'Wilding' as an Invention of the Press." *Race, Gender and Class*, vol. 11, no. 2, 2004, pp. 36–48.

Williams, Raymond. *Television: Technology and Cultural Form*, Routledge, 2003.

Young, L. "Bill Banning iPods in Crosswalks Slated for Albany." 2007. http://wcbstv.com/topstories/local_story_037234835.html. Accessed February 7, 2007.

CONCLUSION

All Users Are Sick: The Normalization of Disease

We are all sick users. This book concludes arguing that the abnormality of disease has become normalized in the sense of being a common and defining aspect of usership. Paradoxically, sickness is deviant, but a typical state, requiring constant surveillance and control, although the responsibility for this varies. Through an extended close reading of four contemporary ergonomics diagrams from the U.S. government's Occupational Safety and Health Administration (OSHA), I argue that the evolving position of the user is a feminized subject with ongoing self-monitoring and adjustment along multiple dimensions, because she is presumed to be already sick. Even if ubiquitous, technopathologies reinforce categories of normal and abnormal, even if abnormal is the norm. Despite moments of derision I suggested previously, almost every member of my focus groups acknowledged experiencing technopathologies directly and in others. These could be quite banal, as when one woman said, "I've experienced carpal tunnel, too, in my wrists from mousing and the computer and headaches, too, from cell phone use too long. … My eyes being tired of just looking at the computer too long." Note how her language places responsibility on herself as user; she does not say "the mouse/cellphone/computer makes me tired." This reflects the trajectory I will discuss in this chapter from technological threat to user responsibility. In the case of computer fatigue, such as eyestrain, it is seen as a source of labor organizing against the health risk from video display terminals in the 1970s,[1] but becomes a list of user-adaptive behaviors by the 1990s. A popular men's magazine included precautions of RSI and other health risks as part of their guide on how to use the Internet.[2] Another article advised, "Don't face a window or bright light ... keep screen straight in front of you ... sit 24 inches away ... use a black and white screen."[3] By the next decade, the user has already taken on this responsibility and blames herself for performing user actions "too long."

Repetitive stress injuries (RSI) are perhaps the most familiar and legitimized type of technopathology, seen in iPod thumb, carpal tunnel from texting, and thumb joint inflammation from early handheld devices such as the Sidekick, Treo, Blackberry, or BlackJack.[4] Of course, this was nothing new: concern around RSI among computer keyboard operators in the mid-1980s reached "hysterical" proportions in Australia,[5] and "telegrapher's thumb" could force an operator to retire from the industry.[6]

I devote this conclusion to such a familiar technopathology to illustrate how disease is becoming common, everyday, and, ultimately, normal. Adopting a different mode of analysis, I narrow my focus to a single type of technopathology, RSI, and examine a single artifact, a set of four ergonomic illustrations, each containing a photograph, diagram, and captions. The set forms the dominant visual of one webpage, "Computer Workstation eTool" of the Safety and Health Topics section of the larger website of the United States Department of Labor, Occupational Safety and Health Administration.[7] I will show how technopathologies continue the individualizing effect of pathological technoculture by creating ongoing cycles of users' self-monitoring and adjustment. The user will be seen here constituted as capable of ongoing processes of self-evaluation with a medical-aesthetic gaze, a necessary process for a user that is feminized, inherently sick, needing discipline, and existing in time. Moreover, we will see how this process varies according to the socio-economic class of users. Although many types of labor can invoke repetitive stress injuries, the white-collar workers using electric communication technologies are addressed differently than manual laborers, privileging certain classes of users and workers while excluding other.

RSI was one of the most familiar and spontaneously recalled technopathologies in my focus groups, and, in line with the individualization effect, understood as necessitating self-adjustment. As one male participant said, "Carpal tunnel. That's basically with the computer. ... So, it's the way your computer is set. If you keep sitting at the same place, eventually you'll get sore. That's why you move the computer or fix the chair." Note that the problem is the user's sitting position and duration, and the way the user sets the computer and chair—not the designs of the computer or chair to begin with or workloads that force the user to be sitting there for so long. Later, when one man was describing the special ergonomic chair he used to stop recurrences of neck soreness from sitting at the computer, a female participant made the individualization even more explicit:

> If you have good posture, no matter what kind of chair you're in, if you've got good posture, you're going to keep your posture up no matter what type of chair you have, but I've experienced that. Sometimes if I'm on the computer for hours at a time, which is most of the time, you know, you tend to lose your posture because you're tired.

As noted previously, this blaming of users, but then acknowledging that they themselves had the same problem, was common throughout my focus groups. In the case of the OSHA illustrations, I will argue that they position the user-viewer as possessing a combination of internalized medical and aesthetic gazes, and participating in ongoing feedback and adaptation, because sickness, while still abnormal, has now become common.

José van Dijck's critical history of social media has documented the process of normalization in terms of how dominant meanings arose to terms such as sharing, liking, and friending during the first 10 years of social media. Drawing on Foucault, she reminds us that "the power of norms, in the area of sociality, is much more influential than the power of law and order."[8] Regulations against texting while driving are one thing, but more powerful control can be achieved through constructing norms—ideals of usership. This can be part of the process Foucault described in that, contra Marx, laboring is not humanity's essence but, instead, complex operations are needed to bind persons to laboring and production.[9]

Ergonomics and the Visual Evaluation of Bodies

Scholars examining ergonomics, the study of human efficiency in workplace environments, often do so within a dualism of individual/organization, particularly from an industrial standpoint of standardizing the individual to the needs of the organization. For example, historian of technology Jennifer Karns Alexander examined the 1929–1932 promotion and display of the

Elmo-workstool, a German product which offered many design features that now define ergonomic seating. However, far from being intended to protect vulnerable workers, Alexander argues this was an attempt to instill industrial regularity and standardization into the varied bodies and movements of workers through "the creation of a pathology of movements that did not fit industrial patterns."[10] A similar process was at work in early twentieth century Germany and Britain with the creation of the pathological category of the "accident prone," individual workers statistically above average in their susceptibility to inadvertent incidents, an attempt to medically plan for and prevent loss of life and productivity in the workplace.[11] Introducing a special issue of the *Journal of Modern European History* on workplace health and safety, editor Paul-André Rosental added a global dimension to these relations of pathologization and standardization, describing how, since the dawn of the twentieth century, legal concepts such as "occupational disease" created tensions in subsequent decades for organizations, such as the International Labor Organization and European Union, as they struggled to transnationally standardize each member nation's efforts at standardizing individual workers.[12]

In short, the individual is a, if not the, problem. The Institute for Ergonomics and Human Factors, an international organization headquartered in the UK, as part of its sixtieth anniversary, conducted a content analysis of articles published in *Ergonomics*, its official journal (and one of the most popular journals from scholarly publishing behemoth Taylor and Francis).[13] It found that the largest category of topics upon which papers had been published—approximately 30%—was Human Characteristics, the individual psychological and physiological factors involved in injuries. This was in contrast to institutional factors, such as Workplace and Equipment Design or Work Design and Organization, which, on average, had a third or less as many publications. Moreover, this was an increasing trend that continues today. In a recent position paper review of research on workplace back injuries in the same journal, a kinesiologist argued that the solution to reducing such injuries was not changes to job design, as these were "impractical" and did not address the "root cause of disabled backs"—but was, instead the workers themselves: "Successful reduction of back injury rates in the future will have to consider 'changing the person to fit the task' or training the way an individual moves."[14] McClusky sees a parallel, opposite move of redistributing social insurance away from workers and toward employers is discussed as part of larger efforts of neoliberalism to replace government security with market risk.[15] In other words, organizational solutions (e.g., buying every worker an ergonomic chair) to organizational problems (worker injuries) run afoul of the individualism of workers. Critical technology scholar Peter D. Schaefer argues that the field of ergonomics as a whole contradictorily seems to offer preventative care and comfort for workers, but this serves as pretense for their exploitation by institutional power.[16] Comforting is a way to preserve worker productivity and labor for extended periods of time, and this also contributes to constitution of worker identities and expectations.

While I appreciate and agree with much of this research, I wish to maintain my focus on people by not examining the user as constituted in opposition to or conflict with organizations (i.e., in a dualism). While, as I will show, organizations are clearly relevant, I want to focus on the user as constituted through their own practices. This is not to say that individuals and organizations are entirely separate, exclusive things. My move here is to elevate analysis of the individual's processes and practices of constituting their usership. Although certainly entangled with processes and practices of organizations, I want to call attention to how it is not a situation solely of institutional disciplining of individual users.

In this chapter's artifact, the OSHA ergonomic diagrams, this is accomplished by evaluative medical looking, seeing bodies with a clinical eye for health and pathology. The user must be able to not only see their own body, but also to see themselves as a subject of medical knowledge. Foucault charts how the perimeters and articulations of knowledge at a particular time and place—the "horizons of the imagination"—map subject-positions available within a discourse to be taken up, and position subjects within that discursive formation. Particularly in the early trio *History of Madness*, *The Birth of the Clinic*, and *The Order of Things*—whose unity Foucault stressed—he explored, respectively, the conceptual models, theoretical networks, and categorical options that made possible institutional structures and arrangements regarding psychopathology, clinical medicine, and natural history. In these archaeologies of varied but related sciences, he delineated components in the process of knowing one's self and place in the world.

Much of these processes involved looking. Martin Jay, in his history of Western ocularcentrism, drew on viewing and viewing relationships in Foucault's work to describe "the complicity of visual domination with the rise of modern medicine."[17] In *The Birth of the Clinic*, Foucault examined a history of medical perception: looking at bodies sick and healthy.[18] He performed a structural analysis of "not only medical discourse, but of the very possibility of discourse about a disease … the conditions of possibility of medical experience in modern times."[19] He described the development of a "medical gaze" during the Enlightenment, a process by which the practice of medicine shifted from a doctor asking a patient, "'What is the matter with you?'" and, consulting his archive of arcane knowledge, choosing the appropriate classification of affliction and remedial treatment, to one of asking, "'Where does it hurt?'" and observing the disease manifest within the patient. This process entailed a bringing into the bright light of the medical gaze "the whole dark underside of disease"—death.[20] That is, modern medicine relied on demonstrating the disease through speaking of one's perception of it, a process based on dissection of bodies of the dead. Disease is understood in terms of death, and the individual is understood in part through disease: "From integration of death into medical thought is born a medicine that is given as a science of the individual. And, generally speaking, the experience of individuality in modern culture is bound up with that of death."[21] The bounded, unitary subject became known through the finitude of mortality, and honed through specific variations of disease. Sickness was no longer a variety of evil things that attacked undifferentiated, innocent victims, it was an encyclopedia of individuals' pathological reactions to their environment: "The space of disease is, without remainder or shift, the very space of the organism. The medicine of diseases has come to an end; there now begins a medicine of pathological reactions."[22]

I have chosen a set of ergonomics diagrams for the clarity with which they suggest Foucault's medical perception: the swollen tissues and sore joints of repetitive stress injuries are clearly "pathological reactions" to the user's environment. While not fatal, I suggest that death still hovers over them. Their forward-looking preventative format of "Do this or else you will suffer" evokes the ultimate "or else"—death. One could protect oneself from outside, invading agents forever; the prevention of decay and injury, however, evokes the body's inherent fragility, vulnerability, and, ultimately, mortality. This universal mortality is, indeed, key to the normalization of sickness.

The diagrams are also visual artifacts: not merely things to be seen but proscriptive guides for how to medically look at users' own bodies. Looking, however, is far from a neutral act. Whereas Foucault theorized power relationships in viewing arrangements such as the panoptic prison, feminist scholars have theorized relations of viewing and spectatorship as not only hierarchical but constitutive of particularly gendered experiences, identities, and subjectivities.[23] Looking not only establishes and maintains power relationships between certain categories, it creates and maintains those very categories as well. Moreover, such relationships can be flexible and protean, as in postfeminist notions of the viewing object empowered or returning the gaze.[24] What I wish to maintain and reiterate here is the notion of looking as relational and evaluative. Viewing and being viewed are not neutral display or inspection but value-based flows of assessment and judgment. Far from neutral, a gaze assesses—even, or perhaps especially, when the subject is the object, as in these diagrams.

The idea of viewing neutrally, specifically in medical looking, is the subject of feminist science and technology scholar José van Dijck's "transparency myth."[25] She argues that medical viewing especially draws on and perpetuates the myth of a transparent body, one that can be clearly and unproblematically known through processes of looking that are not only neutral, but have no effect on the observed. Moreover, such knowledge girds presumptions and ideals of malleability and perfectibility. Through several examples, she challenges and complicates these beliefs, as well as the assumption that visual processes are the best and most reliable way of knowing bodies. When looking at the OPSHA diagrams I will discuss, it is important to resist such ideologies of literal transparency and transparent knowledge in visual artifacts from medicine. As van Dijck and other feminist scholars have explored, the medical gaze constitutes not merely individuals, but gendered individuals, with all their associations and power relations.[26]

All of these conceptualizations of viewing relationships posit separate looking subjects and objects, but what about when these are one and the same? Foucault's subsequent works, such as *Discipline and Punish* and the *History of Sexuality* series, do examine the idea of an internalized, self-monitoring gaze, but does not do so with as much of a specific context of medical perception and disease. Feminist theorizing has also done so, particularly in terms of beauty standards and body norms, in which a feminine subject internalizes the masculine, inspecting gaze ascertained from media images of femininity, and judges her own body on these terms.[27] In the case of gendered diseases, the disease results from various reactions to the ideal image: taking its logics to extremes, protesting it, misunderstanding it. However, what of looking at a picture that is ideal, not as achievement, but as prevention? What if the idealized image does not inspire disease but is explicitly one intended to prevent the viewer from becoming diseased? Moreover, what if the relation to individual subjectivity is even more explicit: it is not merely an ideal image the viewer perceives and then relates to herself—"That is a beautiful woman on the screen. I think that I might be a better, more valued, more powerful woman if I looked more like her"—but it is even more overt in its address for the viewer to interpellate herself into the ideal image—"You must look like this or else you will not be normal. Do this or else you will be worse, sick, failed, devalued, disempowered, and possibly dead."[28]

This is the case with ergonomic diagrams intended to prevent repetitive stress injuries, such as those from working on computers. The individual user is constituted as a technological subject in part through technopathological discourses. In line with Foucault's analysis of medical perception, we see not an invading, evil agent, but the disease "welded to the organism," a literal reaction of an individual's tissues to their environment. Moreover, this medical gaze is clearly internalized: the viewer is put in a clinical position of looking at an image and relating it to themselves. Rather than looking at someone else's sick body while mentally comparing to an ideal of a healthy body, the user-viewer here looks at someone else's healthy body while mentally comparing it to their own. This process is in line with feminist theorizing on beauty and body ideals, yet even more explicit. Instead of the viewer thinking, I should be that image, the image explicitly states, Be this, viewer, or else. Rather than the pathological extreme resulting from the ideal, the ideal is presented to prevent the extreme. The individual in general is constituted through pathological discourse and perception, the technological individual—the user—in specific is constituted through technopathological discourse and perception. Moreover, as with gender, this is a constitutive process of relations, not between men and women, but users and technologies. As with gender, there are hierarchies of value at play: a healthy user is a good user, a productive user, with more social prestige, power, and (ostensibly?) agency than a sick user. As with gender, we also see differences at play: as I will describe, the bodies involved vary in terms of class.

The Temporal User: Interactivity and Feedback

In addition to challenging the transparency of visual artifacts, I also want to explicitly argue against their seemingly synchronic nature. An image suggests a moment. However, through the images in this chapter, I will argue for understanding the user in time, engaged in ongoing processes of evaluative looking and adjustment. In other words, interactively responding to feedback. A technopathology is part of a two-way flow of energy and influence between user and technology, not a single moment of injury or a single preventative adjustment. It is an ongoing process. Carpal tunnel pain does not appear instantly. There are warning signs. It progresses, like many diseases. Pain evaluates, tells users when they are inhabiting their bodies in the wrong positions, actions, shapes, or orientations.

Sarah Banet-Weiser has written about similar interactive feedback and evaluation in terms of technological practices of gender.[29] She has examined how the content of much online interaction in certain media practices of young women and girls tends to be evaluative of the user, forming part of normalizing discourses of ideal femininity and gender. Just as feminist theories of the gaze located viewing relationships in gendered hierarchies of power, here the gaze is expanded to interaction: "looking" becomes "look at me" as girls post videos and photos in anticipation of evaluative responses, ratings, comments, and other feedback to be received and incorporated into future, adjusted, postings and performances. Just as feminist scholars explored the power differential in active and passive looking, here the power differential is examined in the evaluative nature of interactions, recognizing the imperatives for change and conformity its judgments contain. Within pathological contexts, scholars such as Lisa Adkins place such reflexive self-monitoring within the realm of the neoliberal risk culture. Such "surveillance medicine" exemplifies aspects of the individualization effect discussed earlier in which subjects must take responsibility for shaping their own lives and lifestyles.[30] This complements a governmental shift from focusing on expert interventions with what Foucault described as "dangerous individuals" to broader state measures of preventative risk-management at the level of populations.[31]

Or not. The termination, continuation, re-focusing, or returning of the gaze are all decisions, choices, judgments of whether or not the object of our inspection is still worthy of our looking at it. (Here we find ourselves again considering the economics of attention.) Usership is constituted through not only who is looking and how, but also what is looked at, the evaluative information imparted from such looking, what adjustments are made based on that evaluation, and what happens again when looking is repeated, and how these evaluations relate to normative subject positions, such as gendered ones.

But in the case of ergonomic diagrams, the viewer is both. She must view the picture, but also view herself within the picture, and, in an ongoing process of assessing herself in relationship to the picture, sees her body both failing to meet and aspiring towards that ideal. She must reflexively think as both clinician and patient, subject and object, masculine and feminine, valued and devalued. I argue that this process involves not only the internalization of a medical gaze that Foucault and Foucauldean feminists describe—with repetition, specificity, difference, variation, and power relations—but also combined with what I describe as an aesthetic gaze. The user must be not only a doctor and patient, but also an artist and model.

RSI and many similar bodily injuries and mutations can be understood within a classically Marxist perspective as cases in which the user's body bends to accommodate the tool, rather than the tool extending and enhancing the body's productive capabilities.[32] In the interaction between user and tool, the user and her body are more pliable than the device; they are injured by feedback from the device. Users sickened from RSI experience a looping back of mechanical energies into their physical bodies. In carpal tunnel syndrome, the sick user experiences irritation, inflammation, soreness, and sensitivity in the wrists and forearms. This is due to two aspects of the technological interaction. One is shape: the lower edge of the laptop, desk, and/or keyboard presses against the wrists and forearms resting there. This pressure is a message: I am here, we are in contact, you are attempting to occupy the same space as I. The sick user does not heed feedback messages, leaving their arms in place—yet this act of seemingly ignoring the machine is a submission to it. As long as the interaction between user and machine continues, the pressure-message of the machine is stronger than that of the user's flesh. The user's flesh indents, curves, bends. It does not resist the pressure; it receives it; the machine enters the body. The user lets the technology under their skin. Like an over-sensitive person who can't handle criticism, taking it too far, catastrophizing and having an emotional breakdown in response (shades of the excessive sensitivities of neurasthenia and electrosensitivity), the sick user here can't handle the adaptive messages from the technology, and is physically damaged by them. Their bodies surrender rather than adapt, lacking integrity enough to preserve distinction and separateness. Sick users are too fluid, vulnerable, submissive, permeable—bodies almost servile or obsequious in penetrability and accommodation.

This suggests by reduction that the healthy user is instead reasonably accommodating, bending to the technology without losing herself in it. The healthy user heeds feedback messages and lifts her wrists, typing with arms elevated at the appropriate angle. They may purchase a cushion to soften the point of contact. She is pliable, adaptable, and willing, but not to the degree of losing corporeal integrity, of fluidity or deformity. She heeds feedback in smaller, more precise adjustments and refinements rather than absolute submission. As one focus group participant said in response to a discussion of RSI from computers, "Definitely desktop computers [are dangerous] more so than laptops because you have to sit in a certain position. [Another female agreed.] You know, if you have a laptop, you can sit in your bed or get comfortable or you can lay on the ground or whatever." She is the operator who fits her machine well, a keypad operator with calloused thumbs and wrists held at the proper angle, but not broken fingers or carpal tunnel syndrome. She remains her distinct self, therefore knowable and more readily available for quantification and surveillance. The healthy user bends to the machine; the sick user bends over and breaks.

Such interaction between user and machine is also related to the interactivity machines enable between separate users, celebrated in much new-media discourse, when flow between users and communications media is championed for enabling the user to provide feedback, to produce changes to content and/or communicate with the producer or other users. Rather than just sending a one-way message or simply broadcasting a show to an audience, the receiver talks back to sender, the audience tells the producers what they think about a show. Feedback is the person on the end of the telephone call as opposed to a letter; it is the texting hordes of *American Idol* viewers whose democratic revolution storms the Bastille of previous oligopolies of elite talent show judges and A&R executives. As my sarcasm suggests, technological discourse can embrace feedback uncritically. After decades of one-way media broadcasting controlled by powerful elites, the ability of audiences and individuals to not only provide feedback but create their own content has been championed in earlier technologies, such as Super-8 film and cable television public access channels, but exploded with Web 2.0, internet-driven technologies.

But interactivity is at least a two-way flow, with users not only sending but also receiving feedback for them as well. Users can not only produce, they can also return feedback on other's productions and receive feedback on their own in an orgy of comments, ratings, links, stars, forwards, and embeds.

As I have demonstrated here and elsewhere, my analysis casts a narrow eye on the celebratory hype surrounding online feedback and interactivity, recognizing the imperatives for change and conformity its judgments contain.[33] It is a constant stream telling you what you are doing right and wrong. It constitutes an incitement to adaptation. One has agency to ignore it—disable comments, ignore trolls, falsify reviews, rate them as unhelpful—but agency does not mean absence of consequences. As numerous internet marketing books and user guides will tell you, there are real ramifications for not incorporating feedback, in terms of site traffic, search engine rankings, and advertising dollars. Even narrowcasted, niche audiences have to be given what they want—what communication revolution is this, then?

Physical technopathologies are a literalization of this process. Instead of technologies conveying linguistic or numeric messages of judgment on and assessment of our virtual selves (hot or not, popularity rankings, incoming links, favoriting, etc.), here technologies directly impart messages of pain and injury upon our physical selves.

Device modifications could help: you could buy a laptop stand or specially designed keyboard to stave carpal tunnel; iPods added a feature for users to set a maximum volume limit. However, this is a modification of the player, not the rules of the game. Evidencing the normalization of disease, computers and their input devices are not permanently redesigned to be more healthful. The user is expected to adapt their posture and purchase accessories to prevent injury to themselves—or just accept their injured state as part of being a user. Portable music players were not redesigned to be incapable to issuing deafening volumes; their users were expected to set this themselves. Those that don't are bad, stupid users. Moreover, rhetoric of individual liberty reinforce this: People should be free to play volume loud or slouch at a traditional keyboard, just as they should be free to smoke, drink large sodas, or carry guns. Those who get hurt and complain are bad users; those who want to change the devices or their conditions or use are trying to take away your freedom.

Appropriate modulation of interaction requires a user-viewer that merges and internalizes evaluative gazes medical and aesthetic in ongoing processes of assessment and adjustment. She need be able to hold both roles in the viewing relationship, subject and object, but also know how to judge and evaluate those positions, again and again. Sarah Banet-Weiser and Laura Portwood-Stacer, in their analysis of reality-TV programs and beauty pageants, argue that makeover narratives and their "visual display of the plastic, technologized body … privilege a more general cultural narrative about the plasticity of the self," citing similarly plastic selves in Susan Bordo's influential work on body size, eating disorders, and gendered disease.[34] In my specific ergonomic diagrams and technopathological discourse of these generally, we will clearly see "display of the plastic, technologized body." The efficient will be entangled with the attractive, for what are posture diagrams, if not tools in a user's technological makeover?

OSHA's Online Ergonomic Diagrams

The eight images contained within four frames below (Fig. 8.1) are taken from a section of the website of the United States Department of Labor, Occupational Safety and Health Administration. Within the Safety and Health Topics of the website are a series of "eTools," described as "'stand-alone,' interactive, Web-based training tools on occupational safety and health topics. They are highly illustrated and utilize graphical menus."[35] Along with their downloadable versions, PowerPoint presentations on other topics, informational static pages, Expert Advisors (downloadable interactive programs for assisting with compliance in regulated areas such as asbestos standards and fire safety), and eMatrix (a tool for hazard exposure and risk assessment in hurricane response/recovery work), this section of the website provides an extensive discursive formation regarding appropriate laboring subjects.[36] It is literally the State telling people how to work. As such, they contribute to discourse with very real effects, such as lawsuits, fines, and compensation claims.[37] They are part of the production of categories of good and bad work, good and bad workers—including those who use electric communication technologies.

And, in turn, healthy and sick users. The diagrams shown are from the suite of Ergonomics eTools, a microsite on Computer Workstations. Pages address topics such as "Selecting and arranging your workstation components," "Workstation Environment," specific workstation components (telephones, monitors, keyboards, etc.), and the page from which these diagrams are taken, "Good Working Positions." The diagrams are the dominant component of the page, taking up about one screen's worth of a two-screen page. The other half is predominantly thin text, with airy spacing between paragraphs, and consisting mostly of two long bullet lists. I choose to focus on these diagrams for several reasons. One is to apply the ideas of this project to a visual artifact and analysis, and to do so demonstrating the method of an in-depth, detailed analysis of a single artifact group. Second, while I would not go so far as to say this set of diagrams is perfectly representative of all ergonomics diagrams, it does exhibit many conventions of ergonomics diagrams in general and on this website. I also chose this example as one coming from a position of great institutional authority, not to mention accessibility. The site and diagrams are easy to find and view using standard web-browsing technology, and do not[38] employ any high-bandwidth features, such as videos. They are easier to find than, say, an illustration in a manual or textbook, and are easy to view on even a dial-up connection. They are usable both as the page on the live website

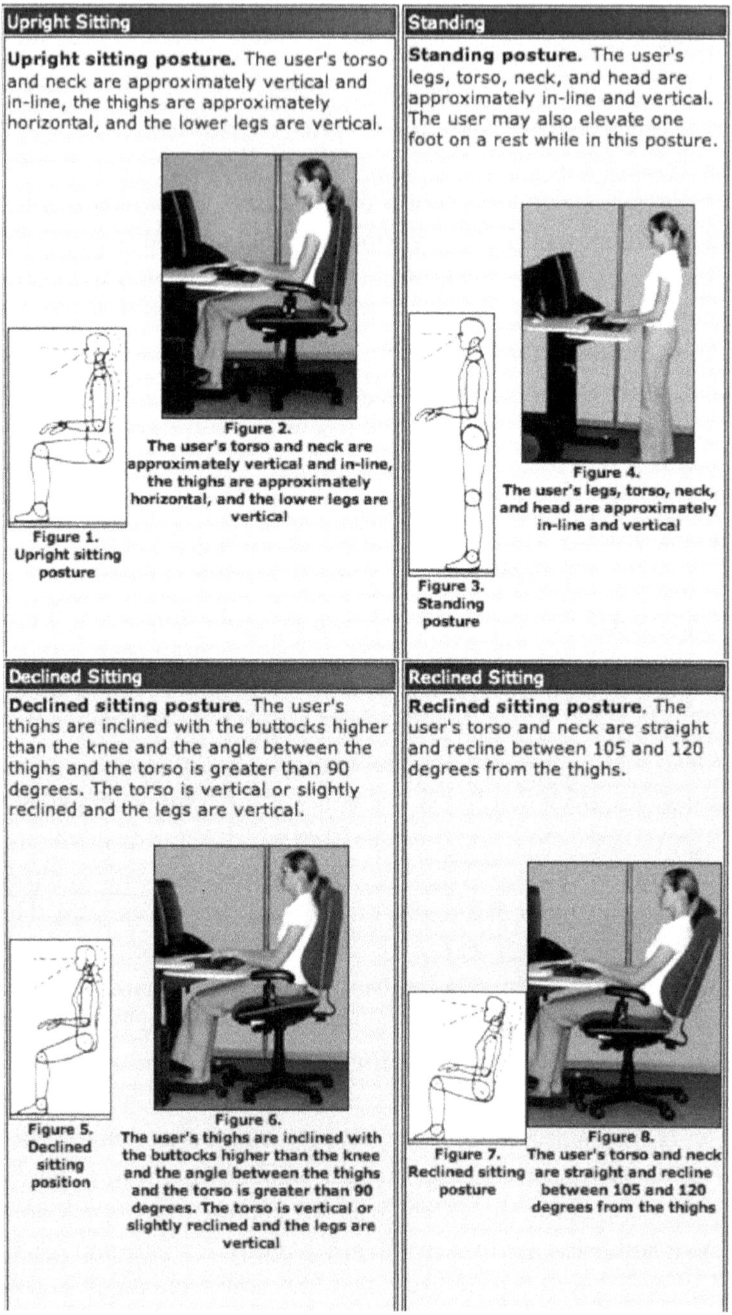

Figure 8.1: "Computer Workstations" (2003, April 23).

Source: OSHA Ergonomic Diagrams. Retrieved Nov. 2, 2009, from http://www.osha.gov/SLTC/etools/computerworkstations/index.html.

and as a downloadable file that could be viewed on an offline computer or presented to groups in a non-networked environment.

I also chose these diagrams because, having employed a sound studies perspective in two chapters, I wished to resist an apparent aural/visual dualism by now employing techniques from visual studies, a field that examines, among other things, what sort of viewing practices and, in turn, viewers, are constituted by visual artifacts. This is not a deterministic perspective of media autonomously shaping consciousness, but one of articulating technological materials and practices that participate in the discursive constitution of subjects, following the ideological currents of the society in which they are embedded. Visual media are arguably apt for examining constitution of subjectivities: Althusser argues that "the structure of all ideology, interpellating individuals as subjects … is specularity."[39] Indeed, several visual scholars have illustrated how analysis of the viewing technologies and practices of looking can suggest social relationships and subjectivities.[40]

I begin by examining certain aesthetic and iconographic elements. Common in ergonomic diagrams are the lines of an angle, often drawn along body parts such as the spine or wrist. Here they are layered over the seat and back of the chair, but clearly indicating the proper-degree angle between thighs and back. The angles shown include an unspecified but visible 90-degree angle, a larger one specified in caption as "greater than 90 degrees," and another described more precisely as "between 105 and 120 degrees." These angled lines presume a viewer with mathematic and anatomical knowledge: she is expected to relate the line to not only the shown body in the picture but her own body, particularly the angled articulation of thigh bones, hips, and spine. The diagram does not show a spine; it is not an anatomical diagram. Arguably, the true curvature of the spine would make such a literal illustration too confusing and seemingly in contrast to the clear, straight angle shown. The viewer then is expected to relate the angle-line to his or her own anatomies, but not too literally.

The line is also red. It is the only red element to the diagram. It stands out even more given that the diagrams and photographs are mostly neutrals: black, grey, white, and beige, the only exception being the blue header band to each quadrant. Red is a color with a long history as communicative, going back to the iron oxide used in Ice-Age cave paintings. Red is a color used throughout ergonomic diagrams on this site and elsewhere, but judiciously, typically to emphasize important elements: body angles, angles of vision, pain radiating out of a neck, and the angled positions and ranges of motion for arms, elbows, and wrists. Red not only emphasizes these elements but their medical significance: from the Red Cross to barber poles, the color has long been associated with blood and life. Ancient Neanderthals sprinkled red pigment on their dead, it is thought, to restore corpses to a warm, lifelike appearance. In Christian art, red symbolizes Christ's blood shed for our eternal lives, as well as live-giving and victorious Divine love, all providing our eternal rebirth. Rebirth through national revolution is also suggested through red's popular association with socialist and communist movements. Red suggests love, sex, and the procreative act. In alchemy, red represents sulfur, which forms a duality with the color white, which signifies the process of transformation, another kind of rebirth. Other associations of white with death—bones, the pallor of corpses—also explain the red/white duality, also seen in barber poles and the Red Cross. It is also thought that this duality relates to ancient theories of procreation, in which red blood mixes with white sperm to create life, and the two colors coming to represent creation in general. Red signifies both life and threats to life. Communicating dangerous threats to life constitutes warnings and proscriptions: do this or else risk a threat to your life, as in red stop lights and stop signs. Here, the warning is to hold this posture or else. The red here, and in significant elements of other ergonomic illustrations, warns not only that this is what must be done to preserve life, or healthy living, but also suggests the creation of life, the (re)production of healthy users, or technologically laboring subjects.

The gaze here is medical in Foucault's sense as well, but a medical gaze that is internalized and reversed. The red lines of posture are also the lines along where pain occurs. It is the disease made literally visible, as in the process of the medical gaze Foucault describes. This is underscored by occasional red elements elsewhere that directly signify radiating waves of pain. Moreover, the process of associating the patient with abstracted bodies of knowledge, of relating the symptoms to underlying discursive structures of knowledge and categories that constitute disease, is suggested here as well by the multimodal style of communication. Within each box there is a photograph, a diagram, and text. The viewer is expected to relate the isomorphic image of the photograph to the abstracted geometrics of the diagram to the linguistic information of the text. This synthetic, diagnostic mode is one Foucault describes, one difference being, however, that Foucault describes a doctor viewing a patient and comparing them to medical knowledge and images in the doctor's mind. How does this patient deviate from health? What disease does such deviation signify? In contrast, in these diagrams, the same act is performed, but reversed and internalized. The viewer is presented medical knowledge and images of health to view. They are then expected to compare them to deviations, if any, in themselves, in a powerful call to reflexivity. The images are taken from a website: you are literally sitting at a computer while looking at images about how to sit (or stand) at a computer; the urge to compare-and-adjust is nearly irresistible. The patient and doctor are united in this model of medical perception, as they compare the visible knowledge in the diagram before them to the invisible experience in their mind of their own sensations and memories.

The Sick Viewer-User

The viewer here, a user, is sick. The presumption is that they will not heed the early feedback in their bodies warning them of approaching disease, and take appropriate measures to prevent it from getting any worse, otherwise why would such diagrams need to exist? Moreover, the tone and address are directed toward the user, not the manager or trainer of the user. It is not "Be on the lookout for this" but "[You] do this yourself." Surprisingly missing from many ergonomics diagrams I examined were good/bad contrasts: do sit like this, don't sit like that. This suggests that the viewer already may be doing the wrong thing. Better persuasion, identification, or interpellation will be achieved by a presenting a positive, aspirational ideal, even though in terms of pure visual communication the message of right and wrong posture could be communicated more clearly by a juxtaposition.

This user needs disciplining. Overdetermination in general suggests anxiety about these apparently simple modifications and postures, that the viewing subject may be resistant or unresponsive. Within each quadrant, identifiers are unnecessarily repeated. For example, the text in the blue header bar reads, "Upright sitting," immediately below which a bold-face lead-in to the copy reads, "Upright sitting posture," as does the caption to Figure 1 and its mouseover pop-up text (not visible here). Four textual repetitions reinforce both a photographic and diagrammatic depiction of upright sitting. Despite the ideology of isomorphic images these diagrams seemingly operate within, the user apparently is not expected to "get it" from images alone. This suggests the user here is conceived as a sick user, someone who does not heed warning messages. Similarly, you can't just tell them something once, you have to tell them multiple times in multiple ways. As with the poor prioritization of channels of information seen with headphone wearers, they need direction, focusing of attention, disciplining of distraction. As discussed previously, the user is stimulation seeking, which justifies the multiple modes and media employed here (text, photo, diagram, popup, download, repetition, and general superfluity). One participant of my focus groups expressed similar sentiment in describing sufferers of RSI and eyestrain as lazy and too inactive:

FEMALE: Muscular problems in general is just not using your muscles enough. They get like lazy, like how kids sit on Game Boys all day or whatever or computers instead of running around playing like I used to do. So, it's just in terms of not using it, I would call that becoming weaker.

MODERATOR: So, like muscles atrophying and becoming weak because they're not being used as well?

FEMALE: Yeah, they're not being used. …

MALE: Just sitting down is not good for the disks in your spine. Sitting down too much watching TV or being at the computer and driving.

Feminized

The user is also feminized in several ways. First, the user is already sick or potentially sick, and the longstanding associations between disease and femininity have already been discussed. One woman in my focus group suggested this of RSI in general, by connecting them to another form of feminized loneliness and over-consumption of media:

FEMALE 1: People who are on computers all day, literally playing PC games all day, I mean that does take an effect on your shoulders and your necks and yeah, the carpal tunnel syndrome. People would definitely not even have to sit at a desk job, they could stay on the computer all day and play a game and get carpal tunnel. Cell phones—people sometimes—it's just people are on them 24/7. …

FEMALE 2: Could I just add something here? When you're talking about—I don't know a lot of people who play games all day long, but I do remember years ago, I think it was on 60 *Minutes* that there are people, mostly women and some of them are truck drivers and stuff, who will read a Harlequin romance book a day. [Female 1 agrees.] So there are people who are so unhappy with their lives that they are looking for an alternate.

Second, in the diagrams, the model is a woman. While many ergonomic photos and illustrations, at this site and elsewhere, use men or unisex figures, the female model here emphasizes femininity. A female model could suggest encouraging feminine (sick) viewers to identify with a feminine image. Next, the feminized user-viewer is further suggested by the model's gazing at screen, not the keyboard, suggesting touchtyping. Typing is a skill with a long association with women, a feminine gendering that Kathleen F. McConnell argues helped to ultimately reassure users of the machine's faithfulness and reliability.[41] Touchtyping is a particularly feminine skill associated with secretarial work, like shorthand, and often contrasted by masculine over-emphasis on an ability only to "hunt and peck." Moreover, it is arguably a lost or disappearing skill, no longer a standard offering of high schools. Why, in a twenty-first century demonstration of appropriate techniques for using new-media communication technologies, include an illustration that apparently demonstrates the necessity of a disappearing skill? And, if it is intended as touchtyping, why not say first and foremost that touchtyping is a necessary skill for proper posture and computer use? I suggest the relegation of touchtyping to implicit rather than explicit expression suggests a residual gendering of the technology of keyboards and an additional gendering of the viewing subject here as feminine. I'm not suggesting that creators of this diagram explicitly said to each other during production, "Women already know how to touchtype, so they can hold this posture, and we don't have to tell viewers to learn to touchtype." What I am suggesting is that underlying gendered expectations of technology use and the technology user viewing these diagrams form the logic behind suggesting a posture that, objectively, is not possible for most users today, because they cannot touchtype. Indeed, one focus group participant discussed lack of typing skills as an example of how her mother exemplified a pathologically bad user:

> My mother likens herself to the Unabomber, and she can't even type. In fact, every time she needs something typed—I bought her a fax machine, which she can manage. She lives in New Jersey. She will scribble something on paper, fax it to me, then I have to type it on the computer, print it out, and fax it back, and then she cuts off the edge that shows it came by fax. That's how she types.

Furthermore, despite the overdetermined masculine rationality of these many charts—their diagrams, numbers, angles, figures, and perhaps even the word "ergonomics"—we are dealing with a traditionally feminine topic, one that has already been suggested several times in my focus group quotes: posture, or, to put it more in the language of Victorian health and beauty guides for women, comportment. It is a subject stereotypically associated with the nagging female dowager relative: "Grandma used to nag at me / to straighten up my spine," sang Carly Simon on an early hit album.[42] Or, as one focus group participant described:

FEMALE: My mom had always told me as a kid, Don't sit that close to the TV.

MODERATOR: We talked about TV and eyesight, but how would TV affect your posture or your muscles?

FEMALE: Oh, I think just the same. I know me, as a kid, I sat on the floor and watched TV. So, if you don't do a lot of exercises or anything, I think, to help your posture, as kids, you have a tendency to slump over, and I think that would develop spinal problems and shoulder problems for sure.

Submitting to or rebelling against such nags is a common trope in narratives of youthful and adolescent masculinity. Moreover, the subject of posture has a gendered history. A 1869 health and beauty guide for women, written by two male doctors and originally published as *The Laws of Health in the Relation to Human Form*, specifies in an early chapter, "The Human Figure as a Whole," that these issues are more important for women: "What we shall say applies particularly to the female figure. For men and queens, says an old French proverb, have the privilege of being ugly."[43]

Finally, in what is admittedly unique to this series of diagrams but nevertheless quite fascinating, is that the diagram—the anatomical ideal to which the user-viewers are meant to correlate the photographs and compare themselves—grows breasts. Both the numbering of the Figures 1 through 8, and their left-to-right, top-to-bottom arrangement suggest a clear, linear progression and order for viewing the quadrants. In each progressive quadrant, as the viewer and the model acquire more appropriate postures and thereby become better users, the model in the diagram develops progressively larger breasts. Why not put breasts on all four images? This is a somewhat stupefying detail, in which one must take recourse in the reality of a discourse, rather than of its subjects, referents, or creators' intent. I cannot fathom any reason for a designer intentionally doing this, nor is my project here psychoanalytically framed such that I could argue it is some sort of visual Freudian slip. Yet there it is, or perhaps I should say there they are, and they certainly contribute toward a discursive constitution of feminine, diseased users.[44]

Temporal

This progression speaks to the user's temporality. It has the effect of underscoring the process of incorporating and adapting to feedback. Constituting the user in time is necessary to speak to the position not as a pose but a process.

One does not merely adjust one's position, one keeps adjusting one's position. One thinks of such ongoing, internalized disciplinary practices—the constitution of docile bodies—as explicated in Foucault's later work, such as *Discipline and Punish*, but the idea is present in *Clinic* as well: "But Death is also that against which life, in daily practice, comes up against; in it, the living being resolves itself naturally: and disease loses its old status as an accident, and takes on the internal, constant, mobile dimension of the relation between life and death."[45] A healthy posture is not achieved but aspired to and continually attempted as part of the internal, constant, mobile relations of daily well-being. Indeed, movement and posture are among the "disciplinary practices of femininity" Bartky describes, along with things such as body size, dress, and makeup, that constitute gendered versions of Foucault's docile bodies that willingly participate in their own processes of normalization.[46]

Here, discrepancies between photos and diagrams subtly acknowledge that these are perhaps impossible ideals, at the very least gesturing to the temporal, ongoing process of adaptation and adjustment: In the photos of Figures 6 and 8, the model's elbows do not achieve the 90-degree angle of the corresponding Figures 5 and 7 diagrams. The model's head in Figures 4 and 8 tilts forward more than in their corresponding diagrams, which suggest downward vision should be accomplished only by lowering the eyes, not the head. In all photos, the model's arms are angled forward, rather than falling perfectly down the center of the side of her chest, parallel with her torso, as in the diagrams. In all photos, each arm and leg is not aligned with the other; the diagrams show pairs of arms and legs so perfectly aligned that we can only see the one closest to us, eclipsing its partner perfectly. It is unclear whether this last discrepancy is a deviation or not. It could be a convention of the diagram to show only one leg and arm in perfect profile, an example of a user who is positioned as capable of the complex medical gaze Foucault describes.

Capable of Medical and Aesthetic Perception

Indeed, the user viewing these diagrams is capable of medical perception. One aspect of the development of medical perception Foucault describes is associated with the development of pathological anatomy, and the change in the semiotic function of the symptom-as-sign. In the earlier clinical model of medical perception, a symptom, perceptible from surface observation of the patient, is a sign of a disease: an observable fever signifies a disease, such as Yellow Fever. However, the development of pathological anatomy leads to greater understanding of pathological processes, heretofore invisible, occurring beneath the surface, such as what Foucault calls tissual communication and tissual reactions.[47] Knowledge of these processes leads to a new relationship of symptom and sign; observable symptoms now better understood as relating to a variety of possible diseases. "Now the sign speaks alone," and, although what it says is necessarily true, it is "never quite certain."[48] This new medical gaze, then, is not only surface observation but, in mentally referencing bodies' internal parts and functions, it is also more spatial, integrative, multisensory, and penetrative. The clinical gaze is a more abstract and spatial process, "a massive *geometry*."[49]

The OSHA ergonomic diagrams encourage such a user as the line drawings and red emphases in the photographs reduce the body to key parts, helping enable any viewer to engage in a clinical gaze. They simplify and encourage viewing that is diagnostic, authorizing the viewer, in a sense, to a medical gaze. Moreover, the viewer of these ergonomics illustrations performs a similar type of perception as they progress, as numerically indicated, from each diagram to the subsequent, corresponding photograph. The diagrams are not anatomical medical illustrations, but resemble more than anything the anatomical dolls of figure drawing and illustration. While they do not show internal bones, they are suggested by the dots indicating joints. Their separation of limbs suggests organs, dissection—that bodies are composed of parts. As black and white line drawings, they are surface and yet transparent: we can view the outline of parts overlapping with others. Moreover, relating back to temporality, as van Dijck has argued, ideals of transparency suggest not only the ability to alter bodies, but an imperative for their change and improvement.[50] The viewing user here performs a "massive geometry" in putting flesh on these diagrams, relating them to the photographs, understanding this relationship as a temporal, ongoing process, and then relating that relation to their own bodies, postures and work processes. The viewer does not reduce the visible to an abstraction, but, starting with the abstraction, progressively embodies what is visible.

Finally, these images constitute a user who is not only perceiving with a medical gaze, but with an artistic or aesthetic gaze. What I suggest here is an amendment to Foucault's medical gaze, which, admittedly, is already quite complex, but it lacks an important aesthetic component. While it is synesthetic in the sense of being multi-sensory—a gaze that not only sees but hears, touches, feels, and tastes—his analysis lacks an address of aesthetics in the sense of beauty, of visual standards of good taste. While staging an encounter between Bourdieu and Foucault on the subject of medicine is beyond my scope here, I do wish to suggest here that the viewer of this diagram does think aesthetically, as an artist. First, the resemblance of the diagrams to figure-drawing dolls overtly suggests an artist's gaze. Second, although the style of these illustrations is not that of medical illustration, their intent and function most certainly are: arranging and posing body parts and displaying structured anatomies in the service of understanding disease, both in terms of treatment and prevention. These may not be "medical illustrations," but they most certainly illustrate medicine, and the history of this act is one of tension between empirical science and humanistic art.

Briefly, early human anatomical illustration was based on animal dissection, revered classic templates in medical books, and surface observation. Before the Renaissance, taboos of defiling the dead and dominant models of inherited medical knowledge from authoritative texts prevented dissection and observation.[51] Leonardo da Vinci's fifteenth-century innovations in drawing from dissection and observation inspired the revolution of Andreas Vesalius, who created the first complete and comprehensive observational human anatomy text in 1543. Anatomical illustration after Vesalius was not only scientifically much more accurate, but prettier: in an effort to respect and humanize the dead, they were classically posed, replicating postures from art and daily life (revived recently in the popular *BodyWorlds* touring exhibits). However, this did not last, as, in this tension between art and science, art eventually was purged as unnecessary and unscientific. Sean Smith's history of anatomic illustration posited an early, crucial turning point in severing art from science with the new, strictly scientific approach of Renaissance anatomist Hieronymus Fabricius ab Acquapendente (1533–1619).[52] However, Scottish surgeon and illustrator John Bell was more instrumental in this development, with his uncompromisingly brute illustrations and influential books. In his 1804 *Engravings of the Bones, Muscles, and Joints*, he bemoaned "the subjection of true anatomical drawings to the capricious interference of the artist, whose rule it has too often been to make all beautiful and smooth, leaving no harshness."[53] This reaction against art became the norm such that, by 1926, the eminent American surgeon, Fielding Garrison, in a limited-edition, privately published history of classical anatomic illustration, issued a plea for art's return from banishment, and made clear not only the importance of the medical gaze, but also its aesthetic dimension: "The practitioner [of medicine], … in bedside diagnosis, must cultivate something of the vision and the discriminating intelligence of the artist."[54] He added later, "The real physician, born to the purple, will always be something of an artist in his way of seeing things."[55]

My point is that, whether incorporating or excising the artistic elements of anatomic illustration, the medical gaze must contain enough of an aesthetic component in order to be able to recognize those artistic elements in the first place. While the sober, monochromatic, overdetermined rationality of these ergonomic diagrams seems bereft of artistic elements, I argue their absence is constitutive. The art-shaped hole in these illustrations of medicine is a heavy presence, gestured to by the figure-drawing dolls, that tells the viewer: "Think scientifically [not artistically]." The art-science dualism is central to the history of anatomic illustration, and the elements of a dualism are mutually constitutive.

Gendered and Classed

Much like that of gender. And, indeed, in the previously mentioned Victorian guide to women's posture and health, the chapter on the human figure makes an explicit connection between healthy shape, pose, symmetry, grace, and aesthetics. Healthy norms are rooted in classical art, whose proportions "cannot be transgressed, even in the smallest degree, without offending a practiced eye. … The medical adviser, … we hope, will bear in mind that he has before him in the case of every woman a double duty, first to save her life and health, and secondly to preserve her personal charms."[56] Moreover,

> one of the most common causes of ungracefulness in motion. … lies in diseases peculiar to women. None but the physician knows how frequent these diseases are. None but he fully appreciates what a terrible foe they are to beauty, not to speak of health and happiness. The lady reclining on the fauteuil, and the wash-woman standing at the tub, victims to these distressing maladies, alike reveal in the positions they assume, and in the gait they adopt, an unconscious effort to "save themselves," and to avoid the suffering which an unwary motion or a painful position gives. What a marplot this is to beauty! [57]

Discourses of healthy posture and graceful positioning appeared in my focus groups as well. As one woman said, "There's positions that you can put yourself in with keyboard, like the thing that you put in front of your keyboard and that's on your keyboard. Like you put yourself at a specific … height … and then it just causes less stress in your body." Another described adapting to the ergonomic keyboards as "awkward," suggesting how she felt her own lack of grace. Another woman, although somewhat dubious, associated her sister's pain specifically with "weird posture" rather than merely repetition of movement: "Some people get carpal tunnel, and I know that some people think that's a direct cause of sitting and just moving and having your wrists in a stable place. Like I said, my sister, she has a lot of shoulder and back problems. She has to go to the chiropractor all the time, and they say it's because you sit at a weird posture all day long." Moreover, several participants in their 50s and older repeatedly asserted that RSI was an aging person's disease. Despite younger focus group members discussing their own experiences of RSI, one woman reiterated, "I don't know anybody young with carpal tunnel, actually." The association of RSI with aging, I suggest, reinforces the association of a healthy user possessing the upright posture of youth and graceful aesthetics.

Finally, there is the issue of class. OSHA offered ten ergonomics eTools: Computer Workstations, Electrical Contractors, Baggage Handling, Beverage Delivery, Grocery Warehousing, Healthcare, Poultry Processing, Printing, Sewing, and Sewing [Spanish]. Of these, Computer Workstations, the eTool from which my illustrations come, was the only white-collar profession (with the possible exception of Healthcare's nurses). The other, blue-collar professions at risk for RSI positioned their viewers differently: few drawings, no juxtaposed photos and drawings, no "massive geometry," and an address to managers, not the actual workers.

Of the other eTools, there were no diagrams at all in Electrical Contractors, Baggage Handling, Beverage Delivery, or Grocery Warehousing. The dense and lengthy Poultry Processing eTool—almost 100 separate pages—detailed guidelines for numerous different roles on the factory line, from Lung Vacuumer to Kill Room Attendant. It typically had a single illustrative photograph of a corresponding factory worker, but these were casual snapshots (possibly stills from low-resolution video) not posed models demonstrating postures or techniques. Nor were they juxtaposed with diagrams. In fact, their purpose was unclear, as they were not in any way instructional: both the employer who hires and the worker who fills the position know basically what a person looks like doing that particular job. Many pages were text-only checklists or litanies of bulleted steps and directions. Two diagrams appeared in Healthcare eTool, but were not juxtaposed with matching photographs. There were several monochromatic illustrations, not in the main eTool but when you linked to their Guidelines for Nursing Homes, but no diagrams at all. A separate page of Ergonomic Equipment featured many photographs illustrating ambulatory equipment, such as lift chairs and bathing systems, but these were more like catalog pictures, and there were no similar diagrams for the posture of the healthcare worker. Printing, Sewing, and Sewing [Spanish] had a few diagrams similar to those in Computer Workstations. Printing's Lithography page reproduced my Figure 8.1 photo, but without the corresponding diagram, and changing the red line to several red circles around different body parts. Further down the page was a diagram of appropriate reach distances on a horizontal work area. However, these were the only two on a page with over 50 photographs and graphic illustrations. Flexography and Screenprinting pages both reprinted the same computer image, within a similar context of many photographs. Flexography also included one atypical diagram juxtaposing do/don't do positions in lifting a heavy roll of paper; Screenprinting had no diagrams at all. The pages of the Sewing eTools were similarly mostly photos and illustrative graphics, with a few diagrams included. However, no section had the same preponderance of diagrams as Computer Workstations. Furthermore, none had diagrams juxtaposed with corresponding posed photos, as I described there.

The lack of standardization in visual communication style was surprising given a subject so standardized: the physics of angles of motion, weight, and force; the geometry of positions. The implication seemed to be that the medical gaze was differently classed. The tension between art and science, the internalized adjustments, the mental connections, the spatial geometries, and the self-monitoring were unique to the RSI of users of electric communication technologies. The subjects and objects seemed more distinct, with all the power relations and valuations this implied: the "user" in the Computer Workstations eTool seemed to be self-monitoring; the "employees" in eTools such as Sewing (note the difference in terminology) did not seem to be the intended viewers, as text repeatedly made indications, such as "Train employees to use proper lifting techniques." In contrast, Computer Workstations directly addressed the user: "This is a comfortable working posture in which your joints are naturally aligned." The potential sufferers of computer-related RSI were granted more subjectivity, autonomy, and agency than those who potentially suffered from its mechanical, blue-collar versions. While the disease and its treatment/prevention did not seem to overtly vary along class lines—good users are still pliable but not fluid—the modes of communicating them, and the viewers they constitute, certainly varied. This asks the question, What of the working-class users of ECTs? In what other ways were they left out, invisible, or otherwise unaccounted for? And the white-collar users of ECTs, if they were reflexive and self-disciplining, if the diagrams address them directly and not their managers or employers—does this suggest that employers and managers are exempt from responsibility for their health and safety?

Normalization of the Abnormal

Writing on pathological beauty ideals, body images, eating disorders, and the larger psychic disturbances they affect, Angela McRobbie has argued that ultimately a network of cultural expectations is created in which individual sickness becomes, not a deviation from the norm, but part of a normal, expected part of gendered, sexed, and sexual experience:

> Better to be an ill girl than a girl who gets up out of her sickbed and challenges the power of the heterosexual matrix. Pathology as normality is preferable to a new form of women's movement. ... Can we talk about post-feminist disorders? ... They become the ways in which "we" understand ourselves as women, hence also a way of excluding those women who dis-identify with these normalizing strategies. ... She who suffers (along with her fellow sufferers) is no longer passive, indeed she is expected to be highly active in her struggle to overcome her afflictions. But these pathologies remain part of her make-up, her personal reminders about what it is to be a woman. ... This also deflects attention away from any idea of patriarchal authority.

Although McRobbie was discussing feminine rage, melancholia, and body issues, consider its insight and application to users instead of girls.[58] RSI is a personal problem; changing the larger system is impossible. McRobbie's "patriarchal heterosexual matrix," here becomes the design and deployment of electrical communication technologies. Personalizing RSI as sickness deflects attention from systemic concerns, as gendered afflictions deflect attention from patriarchal authority. Furthermore, users with RSI are similarly active sufferers—especially when it comes to buying and selling protective devices, such as keyboard, pads and wrist guards, seeking treatments and therapies, and producing and consuming media about RSI.

Why is carpal tunnel conceived as individual physical weakness and susceptibility to repetitive stress injuries, rather than a symptom of dangerous, harmful working conditions? Why are surgeries and wrist braces such common treatments for the disease, instead of requiring more rest breaks, offering on-site massages, or providing different input devices, such as voice recognition software? Ultimately, like ill girls, users with RSI are normal. Pathology is part of the makeup of all users.

Dieting is typically considered a normal part of being a girl, even though a diet is a reaction to an abnormal (in the sense of less than ideal) body weight. It is normal, however, in the sense of being commonplace. Similarly, in my focus groups and in individual discussions with people about technopathologies, people who claimed to be sickened by technologies were ridiculed as abnormal: less than ideal, irrational, possessing various character flaws, much in the way obese people are freely ridiculed. And yet, eventually, everyone admitted to having directly experienced some form of technopathology, just as everyone generally admits to some type of problem with body image or size. The abnormal is normal; sickness is common. For example, consider the common placement in public restrooms of a hazardous waste receptacle for used insulin hypodermics.

Pathological Analysis

This book's analysis of pathological technoculture extends efforts to foreground corporeal and material dimensions of technological experience that had been previously neglected in traditions emphasizing ephemeral communication, broad social changes, collapsed time and space, and bodily transcendence. The examples I have presented suggest, instead, ways in which technology has been understood in terms of concrete side effects, individualized manifestations, specific and particular locations and situations, and deep bodily engagement. Moreover, by focusing on sick users, I have attempted to stage an analytic that more thoroughly unites bodies and technologies in a simultaneous conceptual whole, rather than approaching them as separate entities in various combinations or causal relations.

As I have argued throughout, the larger significance of technopathologies lies in the way they shape ideals and expectations of users and usership. The cause-and-effect relationship of sickness perpetuates understanding of technology and organisms as separate distinct things, encouraging a technologically determinist perspective that frames technologies as somehow autonomous and, therefore, outside of human shaping and control. Users are thereby positioned in a passive position. In normalizing usership, technopathological discourse mark divisions of sick and healthy, and thereby good and bad, creating hierarchies of users. This process draws upon stereotypes of existing social hierarchies, aligning sick, "bad" users with socially marginalized categories, such as gender, race, sexuality, and age. This not only lends legitimacy to our understanding of users, but also reinforces existing social hierarchies. Sick, bad users thus are reinforced by the individualization effect. It is easier to conceive of a technopathology as the user's fault and responsibility to prevent or treat—as opposed to involving technology manufacturers, health institutions, government or employers—when that user is constructed and reinforced as a bad, unreliable, degenerate person already. Such persons are more needing of surveillance and control to protect other users from connecting to their sicknesses, or the reproduction and spread of such sick users. Finally, in a perverse contradiction, the abnormal is normalized. Sickness becomes a normal part of technology use in the statistical sense: it is a commonplace, shared, quotidian, frequent, typical occurrence. Yet, it is still abnormal in the hierarchical sense: it is something bad we must all, then, be on guard to prevent and correct. Disease is everywhere. We should all expect to be potential sufferers. And we are all also our own doctors.

The user is not fixed. Ideals, expectations, and demonizations of users and usership are contingent and in flux. The user is always becoming. I have tried in this book, not to paint a picture of what healthy or sick users "are," but to bring into legibility one dimension of their continual becomings, that of pathology. It is significant because it is not a discrete or isolated thing, but a phenomenon intertwined with bodies, identities, laws, treatments, insurance coverage, and other material and subjective aspects of daily life for many people. As others and I have shown, sickness and pathology are extremely flexible and supple cultural concepts, often contradictory and paradoxical. But, as Sander Gilman has argued, medical discourses have a unique power to confirm, support, persuade, and carry over into other areas of knowledge and life. This very protean and fluid nature of sickness, I suggest, makes it an incredibly useful tool for cultural analysis, as has already been shown, for example, in studies of gender and disease. It is hard to ever pin down the exact "thing" of a disease. Sickness may have a cause, but it ultimately is an experience, and thereby an ongoing process.

Foregrounding cultural phenomena as continuing experiences and process, I suggest, helps them to be seen as contingent and malleable, whether gender[59] or the universe itself.[60] This, turn, helps us all to appreciate human agency to shape, direct, resist, and react to phenomena, even those within a realm as seemingly rational, fixed, and scientific as medical science. (A subsidiary goal of this project has been to suggest how loose and permeable the boundaries of that realm can be.)

In technoculture and other arenas of social life, pathologization shows processes at work and signals the need for attention. Indeed, I suggest that pathologization offers a form of analysis suitable for studying many ongoing process and experiences of different selves, subjectivities, and articulations. I have drawn in this book on work studying such processes in terms of gender, race, sexuality, and nationality. These selves are still in processes of becoming, and always will be, just like the user and many other selves yet to be imagined.

Notes

1. Burgess, *Cellular Phones*.
2. Horkins, "How to Log on the Internet".
3. Young.
4. Menghrajani, "Pain."
5. Hopkins, "Social Recognition."
6. Leonard, *Animals Always*; "This I Believe."
7. "Computer Workstations."
8. Van Dijck, *Culture*, 19.
9. Foucault, *Power*, 78–79.
10. Alexander, "Efficiency and Pathology," 286.
11. Burnham, *Accident Prone*.
12. Rosental, "Introduction."
13. Waterson and Sell, "Recurrent Themes."
14. McGill, "Evolving Ergonomics," 84. A parallel, opposite move of redistributing social insurance away from workers and toward employers is discussed as part of larger efforts of neoliberalism to replace government security with market risk in McCluskey, "Rhetoric of Risk."
15. McCluskey, "Rhetoric of Risk."
16. Schaefer, *Interface Friction*, 49–77.
17. Foucault, *Clinic*, 392.
18. "Clinic" refers to a teaching hospital and the practice of clinical medicine, due to translation from French.
19. Foucault, *Clinic*, xix.
20. Ibid
21. Ibid.
22. Ibid. Foucault suggests that this model only "partially" dominated the twentieth century. However, for various philosophical and methodological reasons, Foucault does not actually analyze the twentieth century in depth. I argue for the continued relevance of "a welding of the disease onto the organism" on the grounds that, while Foucault's model of medical perception has perhaps evolved over the last century, it has not been fundamentally displaced. Moreover, as I have maintained throughout, archaic or residual models of pathology do not just disappear, but linger, echo, and continue to exert cultural relevance.
23. E.g., Berger, *Ways*; Bordo, *Unbearable*; de Lauretis, *Technologies*; Mulvey, "Visual."

24. E.g., Baumgardner and Richards, "Feminism"; Gill, "Postfeminist Media"; Hall and Rodriguez, "Myth"; Holmlund, "Postfeminism"; Sturken and Cartwright, *Practices*.
25. Van Dijck, *Transparent Body*.
26. E.g., L. Cartwright, *Screening*; Hartouni, *Conceptions*; Laqueur, *Making Sex*; Maines, *Technologies*; O'Connor, *Raw Material*; Park, *Secrets of Women*; Schiebinger, "Skeleton"; Van Dijck, *Transparent Body*.
27. Balsamo, *Technologies*; Banet-Weiser, "Girls Rule!"; Banet-Weiser and Portwood-Stacer, "'I Just'"; Bordo, *Unbearable*, "Never Just"; Joseph, "'Tyra Banks'"; McRobbie, *Aftermath*; McRobbie and Garber, "Girls"; Mulvey, "Visual Pleasure."
28. Although Foucault can be quiet on gender, it is implicit with a minor amendment: He examines gazes such as those of scientists, doctors, the law, and law enforcement. These are all positions of authority and power. The crucial and not insignificant amendment feminism makes is acknowledging that, in a patriarchal society, authority and power are associated with normative masculinity and the male sex, therefore Foucault is repeatedly analyzing the male/masculine gazing subjects, even if he does not specify this, and the female/feminine objects of the gaze.
29. Banet-Weiser, "Girltalk.com."
30. Adkins, "Taking," 185.
31. Foucault, "Dangerous," Hooker, "Sanitary."
32. Marx, *Reader*.
33. Scott, "Postfeminist User," "Bubble 2.0."
34. Banet-Weiser and Portwood-Stacer, "'I Just'"; Bordo, *Unbearable*, "Never Just."
35. "Computer Workstations," index.
36. Biedermann, *Dictionary*.
37. Hatzfeld ("Difficulty") documents that European nations were slow to recognize musculoskeletal disorders, such as RSI, as eligible for compensation claims because the causal relation between the disease and work was difficult to prove. As these states progressed toward the European Union, common economic policy, and common healthcare regulations, recognition of the diagnoses increased.
38. Research on medical and forensic anatomic illustration addresses its constitution of racialized and gendered subjects, such as universalizing Caucasian norms (Pugliese, "'Demonstrative Evidence,'" "'Super Visum Corporis'"), the "secrets of women" symbolizing the secrets of the human body (Park, *Secrets of Women*), and the "one-sex model" of female as a lesser or degenerate version of male (Laqueur, *Making Sex*).
39. Althusser, "Ideology," 322.
40. E.g., Crary, *Techniques*; Smith, *Photography*.
41. McConnell, "Profound Sound."
42. Brackman, "The Carter Family."
43. Brinton and Napheys, *Personal Beauty*, 22.
44. This progression toward increasing feminization complicates some gender studies of technology scholarship, which explores the masculine appropriation of mechanics, technology and, most importantly technological competencies, such as the precise motor movements of workers at computers (e.g., Morse, "Virtually Female"; Oldenziel, *Making Technology*). While I agree with work on the masculinizing of technology, I do not think it contradicts such work

to say that the user is also feminized. To consider masculine and feminine as absolute, exclusive categories perpetuates thinking of genders as discrete things or objects, rather than processes and relations. Rather, I argue for femininity *as a normalized component of masculinity*, in a sense similar to McRobbie's (*Aftermath*) notion of the normalization of the abnormal. Diseases, such as carpal tunnel, are a normalized part of technological competency. Therefore the feminized, diseased state is a component of idealized masculine technological usership. One gender can incorporate rather the other because they are in relation, not exclusion. For another example, see the research of Steven Zeeland on male-male sex among military men who do not identify as gay. In particular, in *The Masculine Marine: Homoeroticism in the U.S. Marine Corps*, he documents how "bottoming"—taking the receptive, apparently feminine, position in male-male anal sex—is conceived of *as masculine* because it demonstrates the bottom's ability to endure pain: "Pain is just weakness leaving the body."

45. Foucault, *Clinic*, 155.
46. Bartky, "Foucault," 65.
47. Foucault, *Clinic*, 149, 159.
48. Ibid. 160.
49. Ibid. 159, emphasis mine.
50. Van Dijck, *Transparent Body*.
51. Park (*Secrets of Women*) has identified exceptions in private autopsies and holy reliquaries, where female bodies were dissected and observed between the 13th and 16th centuries. She argues the visible female body played a crucial role in the development and display of medical knowledge.
52. Smith, *Photography*.
53. Bell, *Engravings*, vii.
54. Garrison, *Principles*, 9.
55. Ibid. 56.
56. Brinton and Napheys, *Beauty*, 21, 35.
57. Ibid. 35.
58. McRobbie, *Aftermath*, 96–98.
59. Butler, *Trouble*, *Undoing*.
60. Barad, *Meeting*.

References

Adkins, Lisa. "Taking the HIV Test: Self-Surveillance and the Making of Heterosexuality." In *Contagion: Epidemics, History and Culture from Smallpox to Anthrax*, edited by Alison Bashford and Claire Hooker, Pluto Press, 2002.

Alexander, Jennifer Karnes. "Efficiency and Pathology Mechanical Discipline and Efficient Worker Seating in Germany, 1929–1932." *Technology and Culture*, vol. 47, 2006, pp. 286–310.

Althusser, Louis. "Ideology and Ideological State Apparatuses (Notes towards an Investigation)." In *Visual Culture: The Reader*, edited by J. Evans and Stuart Hall, Sage, 2005.

Balsamo, Anne. *Technologies of the Gendered Body: Reading Cyborg Women*, Duke University Press, 1996.

Banet-Weiser, Sarah. "Girls Rule! Gender, Feminism, and Nickelodeon." *Critical Studies in Media Communication*, vol. 21, no. 2, 2004, pp. 119–139.

Banet-Weiser, Sarah. "Girltalk.com: Technology and the Gendered Imagination." In *Mediated Girlhood: New Explorations of Girls' Media Culture*, edited by Mary Celeste Kearney, Peter Lang, 2011.

Banet-Weiser, Sarah, and Laura Portwood-Stacer. "'I Just Want to be Me Again!': Beauty Pageants, Reality Television and Post-Feminism." *Feminist Theory*, vol. 7, no. 2, 2006, pp. 255–272.

Barad, Karen. *Meeting the Universe Halfway: Quantum Physics and the Entanglement of Matter and Meaning*, Duke University Press, 2007.

Bartky, Sandra Lee. "Foucault, Femininity, and the Modernization of Patriarchal Power." In *Feminism and Foucault: Reflections on Resistance*, edited by I. Diamond and L. Quinby, Northeastern University Press, 1988.

Baumgardner, Jennifer, and A. Richards. "Feminism and Femininity: Or How We Learned to Stop Worrying and Love the Thong." In *All about the Girl: Culture, Power and Identity*, edited by A. Harris, Routledge, 2004.

Bell, J. *Engravings of the Bones, Muscles, and Joints, Illustrating the First Volume of the Anatomy of the Human Body*. Second edition. T. N. Longman and O. Hees, Paternoster-Row, and T. Cadell and W. Davies, Strand, UK, 1804.

Berger, John. *Ways of Seeing*, Penguin, 1977.

Biedermann, H. *Dictionary of Symbolism: Cultural Icons and the Meanings Behind Them*. Facts on File, 1992.

Bordo, Susan. "Never Just Pictures." In *The Feminism and Visual Culture Reader*, edited by A. Jones, Routledge, 2003.

Bordo, Susan. *Unbearable Weight: Feminism, Western Culture, and the Body*, Tenth Anniversary Ed., University of California Press, 2003.

Brackman, J. "The Carter Family" as recorded by Carly Simon on *No Secrets* [LP phonograph album], Elektra Records, 1972.

Brinton, D. G., and G. Napheys. *Personal Beauty, or the Laws of Health in Relation to the Human Form*, Applewood Books, 1994.

Burgess, Anthony. *Cellular Phones, Public Fears and a Culture of Precaution*, Cambridge University Press, 2004.

Burnham, J. C. *Accident Prone: A History of Technology, Psychology, and Misfits of the Machine Age*, University of Chicago Press, 2009.

Butler, Judith. *Gender Trouble: Feminism and the Subversion of Identity*, Routledge, 1990.

Butler, Judith. *Undoing Gender*, Routledge, 1994.

Cartwright, Lisa. Screening the Body: Tracing Medicine's Visual Culture. Minneapolis: University of Minnesota Press, 1995.

"Computer Workstations." *United States Department of Labor*, 2003. http://www.osha.gov/SLTC/etools/computerworkstations/index.html. Accessed November 2, 2009.

Crary, Jonathan. *Techniques of the Observer: On Vision and Modernity in the Nineteenth Century*, MIT Press, 1990.

de Lauretis, Teresa. *Technologies of Gender: Essays on Theory, Film, and Fiction*, Indiana University Press, 1987.

Foucault, Michel. "About the Concept of the 'Dangerous Individual' in 19th-century Legal Psychiatry." *International Journal of Law and Psychiatry*, vol. 1, 1978, pp. 1–18.

Foucault, Michel. *The Birth of the Clinic: An Archaeology of Medical Perception*, Vintage, 1994.

Foucault, Michel. *Power. The Essential Works of Foucault* 3, The New Press, 2001.

Garrison, Fielding H. *The Principles of Anatomic Illustration before Vesalius: An Inquiry into the Rationale of Artistic Anatomy*. Special edition of 110 copies "printed from type on Kelmscott hand-made paper, with the illustrations on Japanese vellum." #97, signed by the author. Paul B Hoeber, 1926.

Gill, Rosalind. "Postfeminist Media Culture: Elements of a Sensibility." *European Journal of Cultural Studies*, vol. 10, no. 2, 2007, pp. 147–166.

Hall, E. J., and M. S. Rodriguez. "The Myth of Postfeminism." *Gender and Society*, vol. 17, no. 6, 2003, pp. 878–902.

Hartouni, Val. *Cultural Conceptions: On Reproductive Technologies and the Remaking of Life*, University of Minnesota Press, 1997.

Hatzfeld, Nicolas. "The Difficulty in the Recognition of Musculoskeletal Disorders between Transnational Medical Specialists, National Authorities and Social Players." *Journal of Modern European History*, no. 2, 2009, pp. 265–282.

Holmlund, C. "Postfeminism from A to G." *Cinema Journal*, vol. 44, no. 2, 2005, pp. 116–121.

Hooker, Claire. "Sanitary Failure and Risk: Pasteurisation, Immunisation and the Logics of Prevention." In *Contagion: Historical and Cultural Studies*, edited by Alison Bashford and Claire Hooker, Pluto Press, 2002, pp. 129–152.

Hopkins, Andrew. "The Social Recognition of Repetition Strain Injuries: An Australian/American Comparison." *Social Science & Medicine* vol. 30, no. 3, pp. 365–372.

Horkins, Tony. "How to Log on the Internet." *FHM*, 1995, October. Reprinted in *Vice*, https://www.vice.com/en_us/article/dissecting-fhms-seminal-october-1995-feature-how-to-log-on-the-internet-101

Joseph, Ralina L. "'Tyra Banks is Fat', Reading (Post-)Racism and (Post-) Feminism in the New Millennium." *Critical Studies in Media Communication*, vol. 26, no. 3, 2009, pp. 237–254.

Laqueur, Thomas. *Making Sex: Body and Gender from the Greeks to Freud*, Harvard University Press, 1990.

Leonard, Mary Delach. *Animals Always: 100 Years at the Saint Louis Zoo*, University of Missouri Press, 2009.

Maines, Rachel. *The Technology of Orgasm: "Hysteria," the Vibrator, and Women's Sexual Satisfaction*, Johns Hopkins University Press, 1999.

Marx, Karl. "Capital, Volume One". In *The Marx-Engels Reader*, second Edition, edited by R. C. Tucker, W. W. Norton and Company, 1978.

McCluskey, Martha. "Rhetoric of Risk and the Redistribution of Social Insurance." In *Embracing Risk: The Changing Culture of Insurance and Responsibility*, edited by T. Baker and J. Simon, University of Chicago Press, 2001, pp. 146–170.

McConnell, Kathleen F. "The Profound Sound of Ernest Hemingway's Typist: Gendered Type-writing as a Solution to the Problems of Communication." *Communication and Critical/Cultural Studies*, vol. 5, no. 4, 2008, pp. 325–343.

McGill, Stuart M. "Evolving Ergonomics?" *Ergonomics*, vol. 52, no. 1, 2009, pp. 80–86.

McRobbie, Angela. *The Aftermath of Feminism: Gender, Culture and Social Change*, Sage, 2009.

McRobbie, Angela, and J. Garber. "Girls and Subcultures." In *The Subcultures Reader*, edited by K. Gelder and S. Thornton, Routledge, 2005.

Menghrajani, K. "The Pain of Staying Connected: Experts are Seeing an Increase in Stress Injuries from Overuse of Hand-held Gadgets." ABC News. 2007. http://abcnews.go.com/print?id=3331731 (accessed July 19, 2007).

Morse, Margaret. "Virtually Female: Body and Code." In *Processed Lives: Gender and Technology in Everyday Life*, edited by Jennifer Terry and Melodie Calvert, Routledge, 1997.

Mulvey, Laura. "Visual Pleasure and Narrative Cinema." *Screen*, vol. 16, no. 3, 1975, pp. 6–18.

O'Connor, Erin. *Raw Material: Producing Pathology in Victorian Culture*, Duke University Press, 2000.

Oldenziel, Ruth. *Making Technology Masculine: Men, Women, and Modern Machines in America, 1870–1945*, University of Chicago Press, 1999.

Park, Katherine. *Secrets of Women: Gender, Generation, and the Origins of Human Dissection*, Zone Books, 2006.

Pugliese, Joseph. "'Demonstrative Evidence': A Genealogy of the Racial Iconography of Forensic Art and Illustration." *Law and Critique*, vol. 15, 2005, pp. 287–320.

Pugliese, Joseph. "'Super Visum Corporis': Visuality, Race, Narrativity and the Body of Forensic Pathology." *Law and Literature*, vol. 14, no. 2, 2002, pp. 367–396.

Rosental, Paul-André. "Introduction: Health and Safety at Work. An Issue in Transnational History." *Journal of Modern European History*, vol. 7, no 2, 2009, pp. 169–173.

Schaefer, Peter D. *Interface Friction: The Politics of Participation and the Spaces between Users and New Media*. Dissertation, The University of Iowa, 2008.

Schiebinger, L. "Skeletons in the Closet: The First Illustrations of the Female Skeleton in Eighteenth-century Anatomy." In *The Making of the Modern Body: Sexuality and Society in the Nineteenth Century*, edited by C. Gallagher and Thomas Laqueur, University of California Press, 1987.

Scott, D. Travers. "Bubble 2.0: Organized, Online Critique of 'Web 2.0'." *Rocky Mountain Communication Review*, vol. 6, no. 1, 2009, pp. 32–39.

Scott, D. Travers. "The Postfeminist User: Feminism and Media Theory in Two Interactive Media Properties." *Feminist Media Studies*, vol. 10, no. 4, 2010, pp. 457–475.

Smith, Shawn M. *Photography on the Color Line: W. E. B. Du Bois, Race, and Visual Culture*, Duke University Press, 2004.

Sturken, Marita, and Lisa Cartwright. *Practices of Looking: An Introduction to Visual Culture*, Oxford University Press, 2001.

"This I Believe." Radio program starring Edward R. Murrow with George P. Vierheller. Originally broadcast August 25, 1952. Tufts Digital Library transcript. Accessed March 23, 2013.

Van Dijck, José. *The Transparent Body: A Cultural Analysis of Medical Imaging*, University of Washington Press, 2005.

Van Dijck, José. *The Culture of Connectivity: A Critical History of Social Media*, Oxford University Press, 2013.

Waterson, Patrick, and Reg Sell. "Recurrent Themes and Developments in the History of the Ergonomics Society." *Ergonomics*, vol. 49, no. 8, 2006, pp. 743–799.

Young, L. "Bill Banning iPods in Crosswalks Slated for Albany." 2007. http://wcbstv.com/topstories/local_story_037234835.html (accessed February 7, 2007).

Zeeland, Steven. *The Masculine Marine: Homoeroticism in the U.S. Marine Corps*, Haworth Press, 1996.

INDEX

Illustrations in this index are indicated by an italicized page number.

A

B

C

D

E

F

G

L

M

N

O

P

R

S

T

U

V

W

X

Y

Z

www.ingramcontent.com/pod-product-compliance
Ingram Content Group UK Ltd.
Pitfield, Milton Keynes, MK11 3LW, UK
UKHW020201290926
15471UKWH00005B/995

9 781433 148453